I0051231

TRANSPHARMATION

How to embrace technology to build a smarter, more
successful 21st century pharmacy before it's too late

ROBERT SZTAR

I dedicate this book to pharmacy owners in Australia who have dedicated their lives and careers to the continued growth of the pharmacy profession, their communities, and more importantly their families.

Especially, those who more recently have seen their pharmacy landscape change rapidly before their eyes, and are struggling to find the time, the resources and the confidence to plan their next move in their pharmacy. I humbly dedicate this book to you.

I hope that the ideas in this book and the accompanying online workbook and resources assist you in creating a healthcare destination for your patients, and your financial freedom or an exit on your own terms.

Thank you…

To my parents Adele and Joe, for providing me with the belief and freedom to choose my own pathway in life, and for supporting me in any way imaginable with every pursuit that I invest my passion in.

To my daughters Sophie and Georgia, I love you more than anything in this world. You are all the inspiration I need to get up early every morning and provide you with a role model who will stop at nothing to ensure that you are given every opportunity and the unconditional support to pursue your passions in life.

To my wife Amanda; ever since we first met we have been helping each other pursue our dreams together. Your love, beauty, humour and optimism has the innate ability to turn my darkest moments of any day into moments filled with renewed optimism and context. Without your unconditional and unwavering support I would not have been able to pursue my passions to their fullest, and for that I am eternally grateful. I will forever love you and dedicate myself to helping make your dreams a reality.

Transpharmation
How to embrace technology to build a smarter, more successful
21st century pharmacy before it's too late

Proudly published in Australia by: Robert Sztar
M: +61 434690579
E: robert@robertsztar.com
Twitter: @robertsztar
Facebook: robert.sztar
LinkedIn: Robert Sztar

www.robertsztar.com

©Robert Sztar 2014
Robert Sztar asserts his moral right to be identified as the author of this book.
All rights reserved. No part of this publication may be reproduced, stored in a
retrieval system or transmitted in any form or by means, electronic, mechanical,
photocopying, recording, or otherwise, without the prior written permission of the
publisher. The only exception is by a reviewer, who may quote short excerpts in a
review.

National Library of Australia Cataloguing-in-publication entry:

Author:	Sztar, Robert.
Title:	Transpharmation / Robert Sztar.
ISBN:	978-0-9924022-0-4 (paperback)
Subjects:	Pharmacy – Technological innovations.
Dewey Number:	615.1

Book production: Michael Hanrahan Publishing (mhpublishing.com.au)
Cover design: Robert Sztar and Peter Reardon
Typesetting: Michael Hanrahan Publishing
Editor: Michael Hanrahan
Printed by: McPherson's Printing

Disclaimer
The material in this publication is of the nature of general comment only, and does
not represent professional advice. It is not intended to provide specific guidance for
particular circumstances and it should not be relied on as the basis for any decision
to take action on any matter which it covers. Readers should obtain professional
advice where appropriate, before making any such decision. To the maximum
extent permitted by law, the author and publisher disclaim all responsibility and
liability to any person, arising directly or indirectly from any person taking or not
taking action based upon the information in this publication.

Contents

CONTENTS

CONTENTS

Getting started:
How to get the most out
of "Transpharmation"

To get the most out of our journey together I have prepared a short video to officially welcome you and help you with your first step.

Welcome video:

www.transpharmationbook.com/welcome

Link to download the FREE Transpharmation App for your smartphone/tablet:

To help you anywhere/anytime you have the time to build your 21st-century pharmacy I have developed this helpful app which provides access to your workbook, our community, the latest podcasts and videos, webinars, and much, much more...all available to you on your favourite device.

www.transpharmationbook.com/app

If you don't have a smartphone or tablet device (or would like to begin your journey on your desktop) you can access your online workbook and our member community here:

www.transpharmationbook.com/desktop

OUR JOURNEY STARTS NOW…

1

Why there has never been a better time for pharmacists to embrace technology

Make no mistake: I'm here to convince you that using technology in your pharmacy each day will make you happier, allow you to make smarter and more effective decisions, wow your patients, and give you more time to enjoy what really matters away from the pharmacy. I will help you shift your current technology paradigm from one that is fearful of the technical aspects and doubtful that it will simplify your current tried-and-tested manual processes to one that clearly understands the technology's purpose and can easily identify areas of your current practice where it can be adopted successfully. No-one knows your pharmacy better than you, and with the world of technology working side by side with you, your future in pharmacy has limitless potential.

The most valuable pharmacy partner you could have

Why do I believe that technology is the most valuable pharmacy partner I could have? Well, technology has been working for me and with me since early childhood. When I was very young my mum needed to buy non-connected power points, double adapters, cables and power boards from the hardware store to help me explore my curiosity towards connecting devices to their power source and each other without electrocuting myself. I had a tireless desire to explore the seemingly endless number of possibilities that could be created using technology.

Pioneering in-car entertainment

My favourite example of this was when I was planning a holiday to the Gold Coast with mum and dad when I was 11, as many Victorian families do. Knowing that the trip would take the best part of two-and-a-half days, I wanted to ensure I had unlimited entertainment options for the journey – I was sure that playing eye spy, navigating and conversation with mum and dad would only stretch so far. I set about packing my newly acquired four-inch Game Gear games console complete with TV tuner attachment, which upon closer inspection offered a unique (at the time) opportunity. It had an audio-visual input socket, which I saw as a path to not only watch TV as an alternative to gazing at the open road and its beautiful coastal landscape, but to watch my collection of VHS videos. Dad had recently been given by a pharmaceutical company a VHS player on which to play the company's promotional material, but he had decided that it held greater value to his son in exploring his quest of innovation. Unknown to him at the time, it was the final piece of the puzzle to bring this solution to life. I was able to plug it into a direct current power source (or rechargeable battery) and the car's cigarette lighter, so its use over a two-and-a-half day trip was unlimited. This conquest taught me that – in partnership with technology – the destination can always remain the same but the journey to arrive there becomes a far smoother, quicker and more enjoyable one.

Throughout my childhood and teenage years I would embark on many similar projects, spending countless hours wiring up electrical, computer and audio-visual components in my bedroom, and any other room that I felt needed technology to make it more useful and effective. Technology and me went hand in hand, and I saw potential everywhere I went to improve existing processes and tasks. But, while entertainment solutions can heighten the enjoyment of our time with family and friends, they don't create or extend the time available to us to spend with them.

The ultimate aim of using technology

During this same period I witnessed the severe deterioration of my grandparents' health, and their subsequent deaths; my Grandpa Sam would long suffer with Parkinson's disease, and his everyday movement was only made possible by pharmacological intervention. Similarly, my Grandpa Leon suffered multiple heart attacks and was only able to successfully recover on a few occasions due to surgical and pharmacological solutions. My Grandmother Clara would die suddenly from an undetected leaky heart valve, ironically on her way to visit Leon recovering from his second heart attack. These experiences taught me that a person's health (both from a treatment and preventative stand-point) when paired with the right technology (which at that time included medicines and procedures) could in practice create or extend our time together with our loved ones.

From this point I knew that a career in the healthcare industry would allow me the opportunity to ensure families everywhere could maximise their time together. With this in mind, I loved spending time with my father Joe in his pharmacies from an early age. He became my mentor and hero. He would selflessly spend six or even seven days a week at work, mostly working 12-hour days in the pharmacies, dedicated to his patients and his busi-nesses. His commitment to the profession, his pharmacies and his family drew my strongest admiration, but nonetheless we all wished he could spend more time with his family.

In my early career I chose clinical pharmacy as the place to expose myself to the coalface of healthcare. While completing my internship at Monash Medical Centre working with my clinical pharmacist preceptors I identified that the invaluable healthcare service we were carrying out was compromised by the bulkiness and abundant nature of our reference material and clinical docu-mentation. So, together with my preceptor at the time – Miranda Ip – and my pharmacy partner – technology – we piloted the use of Palm® Personal Digital Assistants to improve the efficiency of a clinical pharmacist's workflow, and also produced a research paper for the Society of Hospital Pharmacists Australia (SHPA) in 2003. This was my first small-scale project that brought the worlds of pharmacy and technology together.

Following my internship, my wife Amanda and I were lucky enough to continue our career journeys overseas in the United Kingdom, and successfully combined European travel with the experience of working for the UK's leading pharmacy brand, Lloydspharmacy, as Pharmacy Managers for 12 months. This period provided for me key insights into how a pharmacy organisation – irrespective of size – can be successful if their core systems and processes are efficient and leverage key elements of technology.

Upon our return, I was presented with a unique opportunity to further my pharmacy career – together with both Amanda and my father Joe – at Hildebrand's Pharmacy in Frankston, Victoria. The pharmacy is known for catering to the specific needs of an ageing population, with a heavy focus on clinical and medication management services. At Hildebrand's Pharmacy I made it my mission to piece by piece understand the mechanics of how a pharmacy engine is built, operated, improved and maintained.

Learning from experience...

In the quest for improvement, we partnered briefly with an independent buying group in 2009–10 which was aligned with a retail systems provider. Dad had been on this journey with an existing system provider in the pharmacy industry, but this new system as it was presented was far more advanced and offered many time-saving efficiencies, not the least being the ability to forecast stock requirements and auto-replenish stock. We made the decision to be the first pharmacy to pilot the system.

The experience in our 18-month journey, from an outcomes perspective, was a complete disaster, but it provided me with key learnings on how systems are built, integrated, tested, resourced and improved. We would revert to an existing systems provider, knowing its stability and reliability would restore calm to our store operations once again.

How do we do more with less?

Despite the project's failure, I learnt that technology's purpose is to enhance processes built on strong underlying pharmacy principles. This learning would be the seed that would sprout the root foundations of my company, Pharmactive.

Since its establishment, Pharmactive has sought to answer the incessant question being posed to pharmacy owners: "How do we do more with less?" The facts are abundant: operating and occupancy costs have never been higher, patients have never been so value driven or more demanding of personalised service, and pharmacists are being encouraged to develop a greater role in the primary health community. At first glance, the solution would appear to resemble mission impossible, particularly if we follow the traditional, more labour-intensive methods of 20th-century pharmacy. However, the answer is not to abandon our pharmacies in the midst of adversity but to seek the input and unrelenting dedication from an intuitive, more efficient, reliable pharmacy partner in technology.

Since early 2010 I have lived and breathed this reality and seamlessly implemented over 200 pieces of technology into our everyday workflows, as this represented the only viable path (on both financial and time commitment levels) that would allow me to develop and operate two pharmacies, develop and innovate with Pharmactive, build strong clinical relationships in our primary healthcare communities, and more importantly spend time with my wife and two young daughters.

In this book I will help you understand the opportunities for partnership with technology today, tomorrow and into the future. As this is written for pharmacy owners and not for IT professionals technical jargon will be avoided. I will leave you with a clear understanding of the purposes of technology, its functions, and how your pharmacy can best utilise these solutions with a minimum of fuss. To maximise your benefit from this book, all you need is your best knowledge of your pharmacy processes (which you already have), a willingness to pursue your pharmacy's unlimited potential, and a desire to maximise the time available in both our own and our patients' lives to freely enjoy however we choose.

2

The world has changed – and so must you

A new industrial revolution

The world we live in at the moment is going through a new industrial revolution. Industries across the globe are being transformed. The first industrial revolution required business owners to own the factories and to have all the infrastructure required to bring a new product to market and to sell the product. Today, the factory and the infrastructure are available to anyone for free or low cost.

Then...

Take the travel industry, for example. Fifteen years ago, when planning a holiday we would all have gone to travel agencies, made an appointment with our adviser, and assembled a mountain of printed brochures to review (which was the only source of information that we would have been able to access regarding the travel options available). We'd go to libraries and borrow books to do our research on our holiday destinations. It would take significant time for a family to plan their holiday. And often, as was in my family, we would go to the same destination over and over again because we did what we knew. It was comfortable, and it was easy. When you're away, such as on the Gold Coast, if you had a good experience over two weeks, you would just rebook the same thing, because it was so much easier to say, "Let's do that again, because I have gone to all that effort."

Now...

What happens today when you want to plan a holiday? At the moment inspiration strikes that you want to travel, you can pull your smartphone out of your pocket and have in your hand hundreds of thousands of references, reviews and information on any destination conceivable. If you overhear someone talking on a train about how they have had a fantastic time in Brazil, and you think, "There's a World Cup coming up this year. Let's go to Brazil. What do we need to do?", the information is at your fingertips. Where are the best places to stay? Where in general have people had the best experiences? You have the ability to instantly acquire the knowledge needed to tell you whether there are any restrictions or any issues. You can find all of that out on your smartphone, still sitting on the train. There are sites such as Yelp, Wotif and recommendation sites that can tell you all of this information – and you still haven't moved from the spot that you started this exploration.

The ease of purchasing today

It is an amazing world that we live in today. You can now decide, perhaps over a cup of coffee, that you want to travel overseas. From the same smartphone that you have used to access the information you can book the plane tickets, you can book your hotel or find an exclusive deal. Perhaps you may have been sent one by your favourite travel partner, such as Flight Centre or Expedia. You can do all of that in a lunch break or over a coffee. The speed is remarkable, and you don't require someone to translate it for you.

But the more things change...

But underlying all of this – has the process changed? No, it has not. You still need to find out information. You still need to book your ticket. You still need to book your accommodation. You still need to get on a plane and visit the location. We don't yet have a transporter that can take you from your doorstep to anywhere in the world. (Whether that will happen one day is debatable, but it probably will.)

Technology has levelled the playing field in the travel industry. It is now accessible to everyone. Bill Gates's slogan at Microsoft, from the beginning, was, "A computer on every desk and in every home." But because the only computers around at that stage cost the customer several hundred thousand dollars, this was deemed laughable. But now, we carry a computer in our pockets and chances are we now own more than one, two or even three internet-connected devices.

Banking

As another example, take the banking industry. Growing up in the 1980s, we would always have to go to the bank, make our deposits, fill out the paper forms, visit the teller, get a balance written into our passbook when we finished at the bank. That process would keep us up to date on our finances:

→ We would know our balance.

→ We would know how much money we had.

→ We would know our interest rate.

→ We would have a personal relationship with our bank manager and perhaps the teller.

The processes and relationships instilled confidence that our money was well managed and accessible. However, should we ever have to withdraw money, we would need to forward plan, perhaps up to a week in advance, and decide exactly how much money we would need. Then credit cards came along, and that solved that problem partially. But if you needed to purchase anything with cash, especially at short notice, you would struggle. The biggest change in the 1990s was the proliferation of ATMs which allowed you to instantly withdraw your money and get an instant and accurate bank balance from anywhere in the world.

In the present day, you can access all that information in the palm of your hand:

→ You can transfer money.

→ You can pay bills.

→ You can know your complete, up-to-date financial picture from one of these marvellous devices called smartphones.

→ You can pay a friend.

→ You can pay a restaurant.

→ You can even now – with solutions like PayPal and Square (that's coming to Australia) – pay directly from your bank account using your smartphone at the retail point of sale.

Who would have thought this was possible? Is anyone missing the bank queues? Is anyone missing the paper involved in managing your finances through a bank? You can even apply for a loan over the phone. You can go to brokerage firms and they can tell you what the best home loan deal is. They can compare 70, 80, 100 lenders for you at the click of a button. The financial services sector has undergone huge transformation, and has driven a purchasing revolution: any item you choose now can be purchased at the moment of inspiration. Purchasing decisions that we would normally have waited days, weeks or months for can be done at the click of a button from a smartphone in our pockets. We can do our grocery shopping and even order a customised pizza. One of the best exponents of this is Domino's; you can order any topping, any combination onto a pizza and you're able to customise that and follow its progress right until it comes out of the oven and into your hands. Simply remarkable.

Taxis

In late 2012, the taxi industry in Sydney changed. It is highly regulated, and the number of taxi licences is restricted, virtually guaranteeing each taxi cab a certain income and exclusivity over people requiring an ad-hoc car transport service. But through the agency of new technology, anyone can be a taxi now. Sign up with Uber or GoCatch and, once you complete your online application process, you can take people out in your car on a schedule that you determine. You can take people around the city and earn money and they can pay you. Uber, for example, can promise in peak periods of time a guaranteed car service, when and where you want it. What taxi service in the world can deliver that? Think of the noise that is coming from the taxi industry right at the moment thanks to Uber. These are companies that are entering our communities without concern for boundaries, without any hang-ups about "how we've done things" over the years. Innovative companies seeing problems and not getting in their own way to go out and solve them.

Newsagencies

In the past, to receive news publications and updated special-ised information (magazines), you would always have to go to a newsagency which was exclusive in its territory and licence. Now they've been deregulated. This is a key example of an industry that will no doubt perish in the coming years unless they inno-vate considerably. They're irrelevant. You can now access all the news you desire for FREE. Heard of Twitter? You can now follow any person, any journalist, any company, any media source, any brand, for free. Unlike traditional media that only talks to you once a day (newspapers, news programs on TV, etc.), they will talk to you as often as you like and as often as they like. How many times have you ever picked up a newspaper over the years and that newspaper's reporters/editors gave you the opportunity to instantly talk back to them? To open a conversation. You now have the opportunity of being able to pick up your smartphone, read a little 140-character snippet that grabs your attention, read the article, and if you choose, you can engage with the source, and more importantly engage with the person.

What a world we live in where we can talk directly to these companies, talk to people. In the past, businesses were often classified as "business to business" (B2B) or "business to consumer" (B2C); we are now in the age of people talking to people. Some people say technology promotes anti-social behaviour, and makes connecting with each other more challenging, but it is actually bringing us together in more remarkable ways every day.

Starting a business today

All the tools that these businesses are coming into our industries with are transforming our lives, and we all marvel at these young entrepreneurs who have these ideas that just take off and they can earn six- or seven-figure amounts in their twenties. We think that they have an insurmountable advantage as technology wiz kids, but to put it simply they are savvy, not geniuses. They have come into the world we live in and they don't have 20 or 50 years of hang-ups about how these industries have been run. They let go of any history. They don't have boundaries. They don't have fences to jump over. They don't have strings tying them down. They have an idea, they see a problem, they see a need, and they go for it. So, how are they so nimble? How are they able to rise so quickly? How is a company which is essentially a start-up two years ago, like Uber, able to take on taxi industries all around the world that have been well-established for so many years? How is it that they're able to do that?

Getting connected

The tools required have never been more accessible. To get started, most businesses need a phone network. We need to be reached. That's cheap and easy. We need internet. We are now getting the fastest internet speeds in the world, where gigabytes of files can travel around in minutes. We are able to access a library of resources and learn on the fly. You can learn about any person in the world that may interest you. You can watch videos. You can watch TV shows that come from the other side of world. People who are 17 or 18 years old can start businesses in a matter of days. They all have phones. They have internet. They can start a social media page, develop a network or community, have conversations with people who are interested in them, make videos, write blogs.

Blogs

Blogs are like newspapers of today. Anyone can do it. Anyone with an opinion can write an article and it can be read instantaneously. There is no need to wait. There's no editing process. No printing or distribution. People can be as prolific as they choose, and the ideas have never been more free flowing. You have social media: free marketing, which is a great translation for it. But is social media such a new concept for everyone? Haven't we been talking all these years? Hasn't the aim been to talk to customers? To talk to people we are interested in. To talk to family, talk to friends; it's not new. It's a conversation in the 21st century. You can exchange photos, ideas, videos, and collaborate in a matter of seconds. These things don't have to wait. In years gone by, when we came back from a huge holiday trip we would have to get

16

the whole family together for a slide night, video night or photo album night. The people around us weren't able to engage with us until weeks after we had an experience. Now you can bring your whole network with you anywhere you go, if you choose. Video that you may take at an event can be shared instantly. Conferences can be attended without you physically being there. You can be viewing them live on video or following them on Twitter.

The world at your fingertips

You can have a business where people in other countries can do contract work for you. Graphic designers, transcriptionists and virtual assistants can all work for you without ever physically meeting you. All the tools are available. Project management tools, email, the cloud. No longer does any business or person require a high-powered computer to get the best benefits of every possible technology that is available. Accounting software is now in the cloud. Phone services can even be in the cloud. You can have video conferencing from any device, anywhere you choose, as well as learning platforms. It's remarkable.

This is the world we live in, and in it exist new companies in other industries that are not bound by location, by old methods, by old processes. They are seeing problems and they are solving them as quickly and as efficiently as possible.

Why the pharmacy industry must change

Now, in the pharmacy industry, we need to change. Just like travel, taxis and banks, we are the next industry that requires transformation. If we don't transform our businesses, other people will, other pharmacists that may come from other countries, or they may come from within. As pharmacy owners and as leaders in our communities it is time to stake our relevance. I believe that every one of you had a vision of what the pharmacy you were going to own was going to look like when you came out of university, as did I. Is our pharmacy right now what it looked like when we envisaged it? I would say no. But I believe that every one of you deserves the opportunity to build that pharmacy without limits, without boundaries and without the financial constraints that may have held you back until now.

I chose pharmacy as my profession because...

An exercise for you now. I would love you to spend five minutes on this. Make sure that you are away from the kids, away from the dispensary, and away from any distraction. Grab yourself a cup of coffee, a pen and some paper, and cast your mind back to when you first started in pharmacy. Write down, "I chose pharmacy as my profession because…" List as many reasons as you can think of. Do that now.

....

....

....

....

....

....

....

....

....

....

....

Great. I can tell you the reason I got into pharmacy is that when I was growing up I had grandparents, like everyone else. But my grandparents in my living memory were never well. They were always beaten down by chronic disease. My grandfather Sam suffered terribly with cardiovascular and Parkinson's disease. My encounters with him were very limited. His speech was impaired, his movement was extremely restricted, and the time I would spend with my parents in and out of hospitals and nursing homes was more than I could ever have imagined at a young age. In addition, my grandpa Leon had multiple heart attacks, multiple coronary bypasses, and again had limited ability to interact with his grandchildren. My grandmother Clara died suddenly, ironically on the way to come to see Leon after one of his many heart attacks.

Looking back on that, it was the advent of technology through the many medicines and surgical interventions that were able to keep them going. But it couldn't fix them. They eventually died, and by 17 I had no grandparents. So whenever I see our older customers, I see my grandparents, and I want to help these people. I want to make sure that grandchildren everywhere are able to spend quality time with their grandparents in a better shape than I could. That they are able to have the freedom of good health for as long as possible. As pharmacists, we are able to engage with every one of these patients and find the best pharmaceutical and health care that these patients desperately need. They need us to put them first. That's ultimately what I believe we're there for.

That leads to my next question. Are our pharmacies as patient-centric as you think they are? Again spend another five minutes and have a think about it; without limitation, without financial constraint, are you and your pharmacy providing your patients with the best possible care?

Tying in with question one, another little add-on question; when you came out of university, cast your mind back and have a think about: how did you envisage you were going to look after these patients?

....

....

....

....

....

....

....

....

I imagine everyone had ambitions of becoming pharmacy owners. That was the panacea for us. Was it what we hoped to achieve and is it now? Great. Glad we've been able to do that.

But now we need to get out of our own way; we need to recognise our history. But we also need to let go of it. We need to cast away all the barriers and boundaries that are stopping us from progressing today.

Why is this so important? Our patients need us. Our families need us. At the moment we don't have the freedom to be able to deliver the pharmacy that we and our patients desperately need. I believe that when we put our patients' needs first, our businesses will look after themselves. I believe that we will have transformed our industry when our pharmacies become helpful businesses, truly connecting with our patients by helping them achieve the health outcomes that they so desperately seek.

The challenges facing our industry

One of the limitations that we have at the moment, PBS Reform, is a government controlled payment mechanism. We can't change that. The Pharmacy Guild on our behalf can influence it through negotiation in the next guild community pharmacy agreement. So let's not focus our attention on that. Our opinions matter most to our peak bodies and it's important that our feedback reaches them regularly. Our landlords are not currently understanding the challenges our industry is suffering at the moment, particularly those in shopping centres. We need to, where possible, look to relocate away from the centres if they cannot support us on commercial terms. Our banks are getting tougher on pharmacies and perhaps are not as supportive as we would like. Their evaluations of modern pharmacy businesses are very different and their funding models have changed drastically. Our wholesalers are behaving a little bit like banks too; they want more security as their margins plummet through our heavy requirements of price disclosure and PBS reform. Our suppliers have also had to wear this pressure as debt cycles become bigger for them as pharmacies are struggling to meet their payments month to month. In the absence of a patient-centric pharmacy, our customers value only what's relevant to them, and in a tough economic climate – kick-started by the global financial crisis – this has been price. So, we've seen a proliferation of discount pharmacies, with customers flocking to them because of a clear value proposition of the lowest price with no substantial differences in customer service levels.

So we need to wow our customers and create exceptional experiences that go from start to finish, no short cuts. They want more understanding – there's never been more information readily available to them. They will come into your pharmacy now with an understanding of health conditions and medicines that we've never seen before. So our role changes. We are no longer in a privileged position of possessing the only references of medicine and drug information. Now our customers can access excellent sources themselves. We need to be that trusted adviser and filter. Our society, as we've found, is tech laden. It's social, highly stressed and time constrained, and it's only businesses that are putting customers in the centre, enabling them to work with them anywhere/anytime that they choose, that are doing well and growing at a rapid rate.

Our workforce has changed. The cost of employment has never been higher. The red tape in documentation for employing people has never been greater. The penalties to employers surrounding performance management and redundancies have never been tougher. But there are now tools available – at less than an eighth of the cost of a human resources manager – that can help you manage these.

Drug manufacturers are reducing innovation due to the cost of bringing new drugs to market, which has seen a rise in the last 15 years of evergreening. But now we are starting to see by 2015 that the number of patent expirations and generics – which have traditionally been an area for pharmacies to make good margins – is reducing rapidly.

The threat of supermarket deregulation is still present. However, we would argue that they have deregulated pharmacy by stealth. Retail categories that were exclusive to us are no longer exclusive – health and beauty items such as toothpaste, toothbrushes, shampoos and conditioners. Low-level analgesics like ibuprofen and paracetamol are all now available inside their four walls. In addition, we've seen pharmacies in supermarkets, and the rise of Chemist Warehouse. We also have copycat pharmacists seeing

Chemist Warehouse and wanting to be like them and displaying "Discount Pharmacy!" on their shopfront, thinking this is the best model that they can put to their customers.

Can anyone honestly say to themselves that when they came out of university, they said, "I want to be a discount pharmacist. I want to cut my services. I want to cut prices." I don't think anyone would have. I think if we were cutting services, cutting prices and cutting value to our patients, we would never have passed our pharmacy exams. Our role is to be totally and utterly comprehensive. We are still one of the most trusted professionals in this country, and around the world. I believe that this is possible. And in the 21st century, there has never been a better time for a pharmacy owner to be able to do just that.

We cannot be everything to everyone. Our stores have always represented a space that has more than just what the patient wants. We need to transform our stores too and how they look and how they feel. One only has to cast their mind back to the 1800s, as I did in a recent visit to the historical town of Sovereign Hill. These pharmacies had retail, lots of retail. But the retail items were manufactured on premises. Soaps, creams, anything that the community can benefit from, and because the manufacturing was done on premises, the merchandise always matched the community. There were no large distribution national brands that would want to put their product in every pharmacy in Australia. It just was not possible. Each pharmacy owner listened to their patients and merchandised their pharmacy accordingly.

Over the generations, that all changed. As there was an absence of gift shops, toy shops, photo specialists and lotto agencies, pharmacies just took all that on and said, "We'll do all those things for everyone because we can." Financially the business was always sound. With a strong and seemingly infallible dispensary behind the pharmacy propping it up, a pharmacy could incorporate almost any retail activity into its four walls without risk. But that's gone. We are now seeing pharmacies suffer the ultimate pain and closing their doors voluntarily and involuntarily due to liquidation and administration, or even bankruptcy. We would never have imagined that any pharmacy could ever suffer that fate. But nothing can be taken for granted anymore.

The world has changed, and so must you.

3

Step 1: Education

The four steps of Transpharmation

In my 15 years of experience in community hospital and international practices I've partnered with technology every step of the way. More recently, in community pharmacy I've implemented hundreds of pieces of technology that up until now haven't sat within the pharmacy space. This technology has been taken from other businesses and other services and implemented in the pharmacy environment using the four steps that you are going to learn about in this book. If you follow these four steps to a tee, you will achieve transformation – or what I call *Transpharmation* – in your business.

It's very important that you follow these steps in order as there are hidden elements to this process, and the results of this will astound you. Working through this program, your eyes will open, barriers will fall down, boundaries will no longer exist. You will feel refreshed, confident, and have a new-found motivation for your role as a pharmacy owner in the 21st century.

Nothing like this has been done before. In tying together my experiences over my 15 years and also observing the best practices in and out of our industry, both local and abroad, I found that these four steps will give you your best chance at success. I believe that if you follow these steps you can transform your current business into the one that you dreamed of owning and running when you first came out of university, and you can do it in just 12 months! You do, however, have to do these four steps really, really well, but – as promised – you are going to get a lot of help along the way. Each of these 4 steps comes from my best thinking and my best knowledge. If you follow my instructions and the steps I have laid out for you, you will be successful in transforming your business.

The tools, the knowledge and the insights you need

I suppose you're going to be wondering, is this hard? When am I going to be able to do this? Who's going to help me?

This book gives you the tools, the knowledge and the insights you need to be able to map this out for yourself. You will walk away from completing this book with a clear plan and you will understand your business better than you ever have before. You will be able to instantly recognise opportunities, and act on them. No longer will you be bound by limitations and structures that have held back our colleagues before us.

This is a new age for pharmacy. This is a new age for business. Businesses in all sorts of industries are achieving success and they are innovating in ways that people didn't think possible.

You can too.

Disclaimer: You may be thinking that education will take a lot of time, but if you are going to have that pharmacy that you've always dreamed of, it starts with building knowledge through education. Open your world to everything that's possible and you'll open the door to a genuinely profitable 21st-century pharmacy that is operationally efficient, patient-centric and utilises smart technology.

It all starts with education

The first step is education. Chronologically, our lives begin with education from the early development stages; primary, secondary and tertiary education are all designed to give us the grounding to map out the life and the career of our choice. But this can also be quite prescriptive, and set you on a path that is difficult to divert from. The method of education in this book is based around the concept of applied knowledge, not prescriptive knowledge. When you have this knowledge under your belt you are less likely to trip up on your next step because that next step may have already been taken by someone before you, so you can know exactly how to move forward to guarantee your success.

This first step is the most critical. Without it, the other three steps don't matter. You need to understand the landscape of the world we live in, and particularly that of technology. The education step of Transpharmation is designed to give you the best knowledge available and to help you translate that into understanding every purpose of every piece of technology available and instantly be able to decide if it's relevant to you and your business.

Ignorance and a fear of the unknown will make the next three steps impossible. What you will gain from this step is the knowledge required, and the knowledge will inspire confidence, and when you are confident you know exactly how to move forward. What you will find moving through this program is that every step of the way, even the smallest little step, you will get more and more confident. Your eyes will grow wider and the vision that you may have previously had for you and your business will soon look like it was too small. There is so much opportunity just waiting for you, and there has never been a better time for a pharmacy owner to take full advantage of this.

To go through an education course that is unstructured is often the best method because you will instantly find areas that are of interest to you. Some things will make sense to you and some things won't. But I encourage you to listen and read as much as possible to be able to embrace as much as possible. This is what I have done over the years.

So let's begin your education – and start your *Transpharmation*.

Tip: To gain the best benefit of Step 1: Education, create a technology knowledge base for you to clip articles, note ideas, capture photos, or any information that sparks your interest as you work through this chapter. You will then have an invaluable reference to help you complete the upcoming steps, and beyond.

Google

The place to start is Google. Google's mission is to organise the world's information, and make it universally accessible and useful. In the past, libraries were our main source of information; it was clumsy and it was time-consuming to get the information you needed. It certainly wasn't dynamic.

Google is the best place to start when you want to educate yourself on what technology is available. Today, you can find out anything – anything at all – from a simple search. Now, granted, there are a lot of information sources that may not be credible, but that's where your best knowledge as a pharmacy owner and as a business owner comes into play because you're in the best position to filter that: you will know what is credible and what isn't.

So where do you start? You can type in "technology" and a huge world will open up to you. You will see names of magazines, journals, business publications, case studies, blogs and countless other sources of information. You will be bombarded. That's why it is important that you brace yourself as much as you can, so that you are not awestruck by the volume of information.

Trade magazines

To begin to filter these results, start with business technology. Trade magazines within the pharmacy industry have elements of technology, and you'll be able to find these online. I encourage you to read those, both Australian based and abroad.

What will you come across in these magazines? There will be useful articles. There will be case studies. You can read those in conjunction with your best knowledge of your industry and without limiting yourself to your current business. Think about what these businesses have been able to achieve by partnering with technology. The case studies will often show the beginning of the business, what tools and technology were used, and what the end result was. From the case studies that are available, you will find that there are many industries making innovative gains on the technology front. Not because they are developing technology, but because of the application and utilisation of it and what it is doing to transform their business. Ultimately, this is the definition of Transpharmation: a marked change in the optimisation of a pharmacy business through the implementation and utilisation of technology.

Podcasts and audiobooks

Podcasts and audiobooks are a good resource as well. You can learn as you drive, complete errands or exercise, which traditionally were activities that couldn't be linked with learning, as that always required you to look directly at a book or screen to read information. It is also a time which we perhaps underutilise currently as productive time to and from our pharmacies each day. There has never been more specialised content available in this format; relevant to anything you may wish to learn about, even pharmacy and technology.

Learn from other industries

There are often parallels that you can draw from other industries; the hospitality industry, for example. Restaurants and cafes are very similar to pharmacies. They thrive on vibrant, dynamic customer service. If you go into a cafe and order a cup of coffee and it takes half an hour, you're not coming back. The kitchens in restaurants require dynamic systems, organisation and technology to make their businesses work, as do our dispensaries. There are a lot of parallels if you look outside of the pharmacy walls.

In a restaurant, orders need to be placed. These orders are often documented on pieces of paper, although more and more this is being done electronically. That order needs to be sent to the kitchen for fulfilment. The order needs to stay in motion as much as possible to be able to deliver back to the customer their food in the shortest space of time. They're under a lot of pressure, because having a hungry customer is one thing but having an impatient customer is another. One could argue that the stress in a restaurant or in a cafe is often higher than in a pharmacy, and technology has evolved rapidly in that space.

Earlier I mentioned Domino's pizza, which now provides the ability to customise a pizza from a mobile application. Look up Domino's. Research Domino's. Look at how they do it; how they deliver, how they developed their systems and how they implemented them by partnering with technology. How do restaurants do it? How do cafes do it? These are very closely aligned businesses to ours, and they can provide some valuable insights as far as case studies are concerned.

Our industry does not have a massive ecosystem of technology products. It is quite limited. The majority of our pharmacies may hold a point-of-sale, a dispensing computer, and probably a few other things that we will cover in the next chapter. This fact makes it more essential for you to look at what these other businesses are doing.

I would also encourage you to look at what other healthcare professionals are doing. What nurses are doing. What doctors are doing. What physiotherapists are doing. There are some remarkable stories to consider. There are doctor consultations that are occurring online, psychologist consultations occurring online, physiotherapist consultations occurring online. How could pharmacy best utilise an online presence beyond product sales? I wonder – as will you – as you look through the possibilities. These applications and uses can often be viewed on practitioner websites, videos and case studies (often presented as "White Papers").

A very valuable piece of education is surrounding yourself with people from other industries who are embracing technology. A lot of us pharmacists tend to hang around the same pharmacy owners and pharmacists that we are accustomed to, and we are driven by what everyone else is doing. But because our industry is lagging behind a lot of these other industries that have been transformed by the injection and adoption of technology into everyday business, we need to be speaking to them. Within your network of friends and family there could be a useful connection, even on social media; make it a point to contact them and find our what they are doing in their business with technology. You may surprise yourself – you'll get some great ideas as to how you can go about doing that for your business.

Look at the brands and companies that you identify with, things that you like buying and visiting yourself. Could any of their technology and their processes around technology be injected into your business? It's all possible and available at a click of a button. At zero or very little cost you can gather information about just about any piece of technology that you may want to implement in your pharmacy.

Learn from other countries

This is why I say start with Google, because you will be able to start with a broad search and narrow it down from there. You will be able to tap into the global markets such as the US and look at what companies such as Walgreens and CVS are doing. These are the two biggest exponents of pharmacy and technology in the world. I encourage you to study them. Look at the YouTube videos of some of their flagship stores and how they've embraced technology. Look at what kind of systems they are using. Could these be adapted to your business with just a little customisation? What's making these businesses successful?

Follow and learn from others

As you come across these successful businesses and individuals, follow them. Sounds like stalking? No, it's the 21st-century method of understanding, learning, communicating and interacting. You can follow and talk to these people online on various different platforms. If you can find out something interesting about how they operate, ask them a question. Send them an email. Follow them on Twitter. Send them messages. You'll be surprised how open people are, and are quite generous in sharing knowledge with a fellow business owner.

Twitter

Twitter is a great place to find out the latest news and stay on the pulse of what's going on, as information is updated each second (this is not a misprint). Today this is often where news appears first. Once you find industries, companies or people that interest you, follow them and don't be afraid to ask questions of them too. They'll often write back.

Tip: Twitter enables you to populate "Lists" of people or companies relevant to your interests. This is particularly useful if you – like me – have multiple interests in and out of the pharmacy, and would like to separate or segment them into smaller, more digestible news feeds.

Other social media

Often companies and technology partners will have a digital presence across multiple social media platforms (for example, Facebook, YouTube, blog sites, Pinterest, Instagram, etc.) for various different communications with their clients/customers. It is worth exploring these purposes to discover how this communication strategy could be used in your pharmacy, or how the information conveyed by the companies/technology partners could be relevant to you or your business.

Newsletters

Technology and industry newsletters are a great opportunity as well; you will find subscription sign-up boxes on most company or industry websites. Most are sent to you once – or possibly twice – a week. Read them for a few weeks running. You may find a nugget or two of insight that you could bring into your business. If not, it is very easy to unsubscribe at the footer of the email, to avoid receiving them again in the future.

Tip: You can create an email filter rule to save these newsletters in your knowledge base automatically. You can then easily retrieve and refer to them in your planning for and implementing the next three steps, and throughout your Transpharmation journey.

Visit other stores

To gather ideas for your store, visit other retail stores. Visit clothing stores. Go to your nearest shopping centre and do some visits and follow their customer flow. Go to an Apple store. Watch the customer service and how that is heightened and delivered by technology. It's an absolute must. If you want to see how easy it can be to look after a customer in a first-class manner without the administration getting in the way, check them out. You will be astounded.

You can examine electronic businesses such as JB Hi-Fi or The Good Guys. Master's Hardware are offering "Click and Collect", as are Dick Smith; a huge number of businesses are offering that now. They are able to offer their entire catalogue of products over the internet through mobile devices, allowing their customers to order 24 hours a day, seven days a week. Look at what they are doing and their type of systems.

Often when I go to visit one of these businesses I'll have a chat with one of their assistants to find out a little bit about how their technology works. That allows me to project forward about how things may be used in my business.

You will find that a lot of businesses now have a specialisation, and how the specialisation isn't necessarily hurting their business. Through the agency of technology, they can be found easily on Google, and via a number of other methods as well. It's never been easier to establish yourself in a specialist environment. The tools have never been cheaper and more accessible to everyone. You no longer have to rely on that almighty banner on the street as the only method for people to find you. You can be found by anyone/anywhere online. That is why directories such as the *Yellow Pages* have become obsolete. This is where you need to be looking. How do other businesses get found? Where do they show up? What is their presence on Google? Again, it's a good place to start your education.

Television programs and videos

There are many technology and business TV programs you can watch. Sky News has a great program on every Wednesday that looks at technology behind business. I encourage you to watch it.

You can research companies that have been well known for their use of technology in their industry; they will come up on Google.

I find that the best method is to follow a trail of dots and crumbs. Through Google and social media, you can find the end point of that trail very quickly. It could be that you visited a store and you looked at a piece of technology. It might be a screen or a computer or a device; you can note the name of the device and you can use Google to find out what's going on with it. You can very easily work out the pathway of how that business is using that piece of technology and what capabilities it has. Remember, Google also give you access to listings from YouTube (the world's second largest search engine), and for those of us who learn best visually, it is a great platform for education.

Trade shows and conferences

There's a huge annual technology conference in Sydney called CeBIT. I attend it every year. On display are pieces of technology being used outside of pharmacy, and often this reveals opportunities in your own businesses.

Remember, the challenges of business and the functions of business don't vary very much from one industry to another. We are running businesses, and we are also running clinical practices. If there is something that can make the business function more efficiently, it enables us to spend more time in our clinical practice with our patients.

Free trials

Cost is not a barrier. I repeat, *cost is not a barrier*. You can adopt so many useful pieces of technology into your business without great expense. When you are doing your research, don't approach your reading and insights with the limitations of budgets and constraints. Remember there are always cut down or minimum viable versions of anything that is extravagant. You can always make a start, and that's what I am recommending you do. Sign up for some free trials of some software that you may want to try. Get a sample product to try. If you're considering new account- ing software and you want to assess it before handing over the money, sign up for a free trial. Have a play with it. They often have dummy sets of data that you can play with. Then you can decide, perhaps together with your business advisors, "Is this right for me?"

The best education I have had with technology is doing exactly this. Sign up for some free trials, some free versions, and you can learn through trial and error. Just put a little bit of data into these things and see what it looks like. Does it work for you? It could be a customer management tool. It could be a tool such as Mail Chimp, which you could use to send out a weekly newsletter to your customers. Try it out. That's the best way to find out more about these products and how they might integrate into your business.

(We will cover this more in the next two steps. Just keep an open mind as to the possibilities that are available.)

Talk to your peers, suppliers and consulting firms

Have a chat to your peers. We do have some great exponents of pharmacy and technology in our ranks. It may be a minority at the moment, but it will eventually become the norm. Talk to those who have successfully implemented technology. Look at other innovative companies.

Talk to suppliers. Talk to technology vendors. Read about their products. Often these companies have what are called case studies, white papers or e-books. Download them. Read them.

Talk to IT consulting firms. You don't need to engage them, but a lot of them have white papers about their strategies and what they think are the next big things coming up. Follow these trends and see if they are relevant for you.

Think outside the box

Think outside the box. Let's consider the iPhone as an example. When Steve Jobs and Apple invented the iPhone, the uses that they saw for that product went only as far as, "We've got an app store. We've got an iTunes store. We can get all of our customers buying things from these stores anywhere at anytime."

But the uses for these devices have far exceeded just consuming products, media and services. We are now seeing iPhones being used as ECG monitors and sleep monitors, to map out our running patterns, to manage our diet, for monitoring vital signs, and to make payments. There are now many more things smartphones are being used for than was ever imagined when they were invented.

Why? Because of developers, and business people, and non-technical people like you and me. I can't code and I probably never will. But we see purposes in technology that make our smartphones more than just a phone, more than just a camera. Look at our own industry – phone cameras are being used for capturing the QR codes that are going to be on all of our prescriptions, so that we can order directly from our smartphones. That is so far out of the realms of what phones were going to be used for in our industry, but it's happening.

Apply that to yourself. When you look at a device like a smartphone, it's not just a phone. There are tools inside these phones that make them factories. You can shoot a video at better quality than the high-definition cameras of 15 years ago. You can record music. You can record an inspiring speech. You can record yourself. You can record podcasts. All of these things can be done by just applying out-of-the-box thinking.

Anything is possible

Take the approach that anything is possible. If you look at a situation that bugs you and you see it being a problem, don't take it for granted that it can't be fixed. If you don't like that patients have to sign receipts with a pen, look into ways that can be overcome with technology. There are businesses doing that. If paperless is your thing and you think, "I don't want a single piece of paper in my businesses," you can do that. If you want to replace all your promotional signs that you have on gondola ends and your windows with digital screens and be able to easily project marketing material across your store and change them easily every week, you can do that. Anything in this day and age is possible, which is why there's never been a better time to build the pharmacy of your dreams.

Online education

If you would like a slightly more structured, yet flexible education there are courses that you can do. These are what's called "webinars" run by technology companies and learning institutions or online courses, available through iTunes university, Udemy and others. If you can think of a skill that you may want or a technology that you may want to learn about, you can do a course on that anywhere, anytime, right from your smartphone. The time that you put into this will determine your success. Knowledge equals power in the new world of business. The more you know, the more you'll be able to transform your business.

The best education you can obtain is the one that works for you. It is so important that we do our education before anything else. It may seem daunting because it's an area that you may be unfamiliar with. But it has never been easier to access everything that we have spoken about and listed here. Have fun with it. It's not structured and regimented like going to school.

This is the new world of business. You will find the more you know, the more your confidence will grow, and the more excited you'll be to move to the next step.

4

Step 2: Discovery

Well done on completing step one! Education is ongoing, but it's great that you now have some baseline information around you, and with that, you can move on to step two.

Step two is discovery. This is where we're going to look at your existing pharmacy (or pharmacies). Using your newly acquired knowledge and armed with the new insights that you discovered in step one, we can now look at your existing pharmacy and discover opportunities to apply some of those insights to transform your existing pharmacy. It's a massive change to consider: that the pharmacy you stand in today could transform into a profitable 21st-century pharmacy. Follow this sequence and you will be able to achieve that. I have no doubt.

Discovering what you have – and what you need

So what do we need to look at in discovery? What do we need to have in front of us to be able to make the right decisions about the new technology that we have just discovered, or technology that we knew existed but we didn't think it could be integrated into pharmacy? Where can we start?

Your operations manual

A good starting point is your operations manual. Don't worry if you don't have one or if you're not quality care accredited. The best method of getting a baseline operations manual under your belt so that you can begin this process is starting with a reference quality care manual. So if you haven't already enrolled in Quality Care, I would encourage you to do so because it will give you some baseline policies and procedures that we will be able to work through in finding opportunities for your pharmacy.

Where this is important is Quality Care will break up your pharmacy into sizable chunks. What we're looking at here is 18 elements that are required to meet basic standards, things such as legal and professional obligations, supply of medicines – you know the list. There are 18 different sections. Even the management of Quality Care, as arduous as it seemed in the past, can be achieved relatively simply by partnering with technology through the agency of Google Apps, which is a suite of products that you would normally have seen from Microsoft which includes the equivalent of Outlook, Word and Excel. And now

Microsoft has Office 365. So now all of the physical documents, policies and procedures that existed in your operations manual can be digital, moving, breathing documents.

Why is that important? Well, when you're in your pharmacy, is it only one person who operates it? Is it the Quality Care coordinator who operates your pharmacy? No. Does every person have access to that Quality Care manual in any area of the pharmacy? Currently probably not. So by using the Quality Care fast track, you're going to be able to get electronic copies of all of your policies and procedures. By injecting those into an online office application such as Google Apps or Office 365, you and your team will be able to maintain and access the files much more easily. You may assign as part of your job descriptions in your pharmacy procedures to specific people who have specific roles – no longer do you need an overriding Quality Care coordinator. Everyone now is collaborating and putting their best thinking into developing your policies and procedures. I think this is a critical step to take in your pharmacy while you are undergoing Transpharmation, because what you may find is that in the policies and procedures you may have currently – or if you are just using the standard policies and procedures – there are steps in these processes, or entire processes, that could be automated or aided by technology. You will need to overlay your efforts with technology in building your business with the purposes of these technologies to be able to do it successfully.

You will not need to rewrite your Quality Care manual. You do not need a magic new 21st-century set of procedures. The principles, fundamental procedures and theories do not change.

As we've already discussed, our sole purpose is to solve problems for our patients. Everything else is a distraction – and the distractions can be minimised by technology. While we require certain levels of process management and administration to run our pharmacies, we need to minimise this impact. That's where technology is most valuable to you, in addition to heightening your patient experiences.

Following the four steps, we will go into detailed cases studies that will show you how this can all be applied.

Let's first focus on what you need to discover about your current business.

The operations manual will allow you to go step by step and find opportunities. You can look through each of those sections and find ways to be able to become more efficient. You could look at issues such as:

→   your business process management
→   your admin structure
→   how paperwork is filed
→   how prescriptions are accepted.

There may be certain levels of technology that can be introduced in these areas. For example, rather than type writing or hand writing prescription receipt dockets, you may decide to do that through your dispensary system now that you're aware that this functionality is available. Or you might use something as simple as a time stamper so that you are able to file the prescriptions in chronological order and have an audit trail for you and your patients.

Your team

You need to be aware of what the current roles are within your team. What are the job descriptions? What are the roles that we have in the pharmacy? You probably need to rank those roles according to what impact they have on your customers and your ability to solve their problems. Any job description you find that has a huge weighting to non-customer-facing and non-customer-engagement activity, I would flag as a massive opportunity for partnership with technology. You want your best people to be able to heighten your customer experience, not have perhaps a bookkeeper working seven days a week, or more likely five days a week, managing your books because you don't have a paperless system. Maybe you use on-premises accounting software that's clumsy and clunky and requires data to be backed up manually and taken from your pharmacy to your accountant's office manually. All of these things add costs and time that could be best spent with your patients.

You have to really scrutinise where technology can play a role in two key areas:

→ how it can minimise the distractions to your customers
→ how it can improve the customer experience and how it can bring you closer to them.

These areas are going to be the focus here.

Your training manuals

You need to know how you are doing your training at the moment, so you need to have your training manuals on hand. You also need to have your product manuals. If we take the rule of "one percenters", each item in these manuals can add up to a lot of time if you can find the opportunity. All of these things that we are talking about can be made paperless, which means that the time to access them is minimised, and your ability to edit them and share them among your team will in turn improve your customers' experience.

Your asset register

You also need to get your asset register. Some of our assets are going to be tech related and some non-tech related, and some may be owned and some may be leased. There may be better technologies available for some investments that you've made. You need to see if there are opportunities to get more value and more return on any current investments that you might have. For example, you may have a multifunction photocopier–printer that you use only for faxes – you don't use the scanning function at all. But that scanning function could be the central point to allow you to begin your paperless journey, which has significant productivity and real cost savings in saving your space.

In one of our pharmacies we had half the storeroom dedicated to storage of tax documents, prescriptions and duplicates that we were required to keep. We don't have room anymore for unprofitable space in our business. Overall our rents have never been higher. We now need to maximise our space available – it can't be taken up by paper and from a retrieval point of view, this is highly inefficient. So, paperless may be an opportunity for you, and you may already have the tools to begin but previously not had the knowledge.

Your company structure

What is your company structure? Are you a single proprietor or do you have multiple partners? What is your role? If you are in partnership, what are your other partners' roles? What are your aims? What are your objectives? What are you looking to achieve both personally and professionally? All these things need to align to create this pharmacy that we all want, and technology can help us on this journey.

You may be an owner who doesn't spend much time in the pharmacy but who would love to be more closely connected to their team and also to the pulse of the business and its customers. The opportunities to introduce things like remote management systems – they sound fancy, but they're quite cheap to implement – may be very useful. They can log into a terminal at any time of the day and be updated with what's going on in the business, plus they could have tools like instant messenger chat so that they are available for questions at any (or a dedicated) time the staff may need them. This can all be achieved without setting foot in the pharmacy, if that's part of your vision.

Such systems can be particularly important in rural areas, where some owners may be metropolitan based and not come out to rural areas that they are connected with. They can connect by activating CCTV cameras they can access from a smartphone, a tablet or computer, and see what's going on in the business at any time. This allows them to stay in touch with the daily operations of the business.

If you do have multiple partners and owners, you can collaborate using video conferencing and other collaborating platforms that enable you all to look at and work on – for example – the same set of reports and have a round table discussion. This could have previously taken weeks to organise in person with everyone's busy schedules.

Your mission statement

What's the company's mission statement? What does your pharmacy stand for? What do you want to be known for? What is the ultimate result that you want your patients to walk away from your pharmacy with? All of these things are connected with the infrastructure and the joinery that we need to put into your 21st-century pharmacy.

Your pitch

Does your company or pharmacy have a pitch? When a new customer comes in or you talk to someone away from your pharmacy and they ask you, "What does your pharmacy do?", what do you tell them? Do you specialise in something? Are you better than anyone else in a particular area? What are your demographics? What's your customer base? Do you have a niche? Do you have a high concentration of diabetics, for example? Are your systems able to upload all your blood glucose meters? Do you have all the cables and all the software you need to do the job properly? That could be extremely useful for both you and your patients.

Your promotional activities

How do you promote your pharmacy and its services and products? It's important to know this. What types of products and what brands do you have? What services do you offer? You must have a products and services list so that this is clearly defined. What are the specific policies and procedures around those, if they aren't already covered in your operations manual?

Your professional services

Professional services are tough to implement right now. Government funding has dried up, but these services have never been more valuable to our customers. To be able to implement them, we need to consider the most efficient way we can operate these services while maximising the pharmacist's time with the patient and minimising any administration and any other associated noise that comes from running such a service. We need to automate all of that so we can maximise our benefit in implementing those services. So, your products and services may be best advertised online so that prospective customers can easily find you. You can't put them all in a *Yellow Pages* ad, but you can have a website and list all of those products and services on it? What do you tell customers about them? Some of those products and services will lend themselves to technology, and this in turn provides you with another opportunity to start transforming your business.

Buying groups and banner groups

Do you operate under a buying group or a banner, and are there any expectations placed on you because of this? Are there any limitations you may have operating under such a model?

Sometimes some of the technology currently available may not be able to be implemented because of some of the limitations that buying groups and banner groups can place on you. If your banner group, for example, has a prescriptive loyalty card system, you may not be able to implement your own. In such circumstances you need to consider the implications of the ownership of your customer database and your ability to access that database. This fact alone will be incredibly valuable to you when planning your marketing initiatives.

Your premises

Is your premises leased or is it owned? This can be important when you consider the developments that have been forged in essential services management. Are you in a position to perhaps put solar panels on your roof? That may save you a lot of money in the long term. Do you have the ability to install LED lighting to replace your fluorescent bulbs, to save energy costs? How is your building wired? Do you have the opportunity to rewire it to put in a centralized UPS (uninterrupted power source) so that you no longer require individual back-up power for each computer? Do you have the ability to have NBN or fibre cable to your premises? This fact could be decisive when looking forward as to how your business may operate in the future; for example, NBN or fibre may be mandatory for you to offer your patients video consultations or for you to offer live collaboration with other health professionals at the point of care with your patients.

Your financials

Look at your company financials, your budgets, your goals. Again, you may think that the biggest limitations for your business come from these documents, but we also need to look at the benefit end of it as well. Your likelihood of financial success in your pharmacy will change dramatically once you commence building a pharmacy that addresses operational inefficiencies, is patient-centric, and utilises smart technology.

For example, when you are reviewing your company financials and planning budgets, you need to consider that it may be forecast that your profitability will drop as a result of further PBS reform. You can simply reduce operating costs by removing a higher-cost team member (for example, a pharmacist) in your business, BUT it is only a once-off improvement to your business's bottom line, and your ability to generate more sales (from the productivity and skills of that pharmacist) has just been handicapped by that decision. So if we are going to run leaner businesses with leaner staffing models, each person and the business itself needs to pack a bigger punch than it ever has before. Technology can help your team punch above their weight. You can still be a small business but you can access the benefits of what big businesses have had for many years, for little or no cost. But you must understand the financials of your business to see where the opportunities are, and plan to act on them as a priority. Remember, any saving you make in operational efficiency goes straight to your bottom line.

Your past decisions

What has your business looked like in the last two to three years? It's been said that our gross profit levels have increased through some heavy generic substitution uptake. How have you used (if any) that additional income? Has it just helped you survive, or have you used it to invest in any particular areas? This business history will tell us about what decisions have been made historically that can be discounted, and those that could be built upon. What type of pharmacy do your competitors operate? Who are they? Are you surrounded by discount pharmacies? How have you responded to discount pharmacies if they are in your area? Most pharmacies will be facing such a challenge. What initiatives have you tried, and what haven't you tried? Have you tried to implement any technology in your business in the past two to three years, and is it still in your business, and what is it doing now? Did it meet your objectives and goals when you implemented it, or why did it fail? This will also be very useful information in planning your transformed business.

SWOT analysis

I'm going to ask you to do a SWOT analysis.

What are the Strengths in your business? What are the Weaknesses? What are the Opportunities that you see, and the Threats? I ask you to do this after you have done step one because I believe that you are going to find so many opportunities that you have never thought of until now. Put them all down, irrespective of any perceived costs or limitations. It can all be achieved one step at a time.

Your ideal pharmacy

What are your goals from a personal and professional perspective? What would you like to be doing in pharmacy if you had your choice, without restriction? Think in terms of resources, functionality, and the impact that you have on your patients. Don't be limited by past successes or failures. Everything is possible.

Your existing pharmacy

Create a scale map of the layout of your store. This will give you an indication as to what technologies can be used in what areas of your business, using existing infrastructure. This could be something as simple as knowing where your power points are or where your data points are. One of your first priorities in building this pharmacy is trying to reuse as much as possible of your existing setup, without incurring any unnecessary additional costs. Using the map of your store and where your services and counters are now will be central to the first steps of implementation.

You should also take photos of your store. It's going to be great to look at these photos one day and think, "That was my pharmacy

12 months ago and look at it now." There may not be many visual changes; you have to consider that. This would not be viewed as a failure, because a lot of the benefits of technology can be seen and demonstrable without your store needing to actually look visually any different at all.

Your customer traffic

What's your customer traffic like? Have you ever put in a customer traffic counter? What are the common walking patterns of your pharmacy? This could give you some opportunities to be able to look at better methods of marketing to your customers in the store. Where are the high volume areas? Where are the areas of opportunity? Where *aren't* your customers going? What products and services do you focus on in those low customer areas? Do you have a marketing plan of activities and promotions? What have you done in the past? Have you run events in the pharmacy before? How have you promoted those? How did the events go? What was the attendance? Who were the people that attended? Do you have a list of those people? You may have gone out to your local nursing home and done some speaking there. What was the result of that? What were you aiming to achieve?

You may wonder why is this relevant to technology. It is 100% relevant because it will provide you with opportunities to improve. You may – as a result of looking at your marketing plan – think, "Well, we ran flu vaccines in our pharmacy. We did blood pressure checks, and to promote the event(s) we sent out fliers to people's homes. We did a letterbox drop. We put fliers up in the window, and we got 100 people over the course of a week."

But using this new knowledge that you now have from step one, you may decide, "We don't have a social media presence. Let's build one. Let's invite all of our customers to join." Social media, particularly Facebook, is seeing its largest demographic of growth in baby boomers (45 to 64) which represents a large proportion of our highest frequency customers. We simply can't afford to discount their use and participation in social media. It has social connective benefits that even the Facebook founders never imagined. It now connects grandparents with their grandkids every day. It connects families together over holiday periods – they can share photos and communicate while away, when previously they would need to have waited until they arrived home. So you may choose to market these events via social media next time instead of old tried and tested campaigns such as local newspaper advertising, which can be more costly and less targeted. Having this information provides you with a baseline to improve upon.

Your digital assets

In your marketing initiatives, have you developed digital assets? Do you have a logo? Do you have a brochure for your business? Have you created any documents that your patients can download from your website or social media pages? (For example, patient information sheets.) Do you have any digital images of your business that appear online? (For example, Yellow Pages, Website, other directories.) How old are they? It is important that you keep track of these as it's important that your business is seen digitally the same way as it appears physically (for example, if you have re-branded or changed your signage). If you don't already have a website or a digital presence these assets are going to be required to establish an effective and accurate digital presence for your pharmacy.

Your digital activity

What do you do digitally in your business? Do you have a website? What does it look like? When was it last updated? Do you have a strategy around how it looks? I believe that a website should reflect your store environment as it is a digital representation of your entire organisation. So, if you have windows for promotion in your pharmacy and you rotate those every two weeks, it is good practice to ensure that you rotate your front landing page or banner ads every two weeks as well.

Make sure you keep things interesting for your customers. There is nothing worse than coming into a physical store and thinking, "Yep, I saw that last time when I was here eight weeks ago. It hasn't moved. They are still doing that." You have to keep it fresh and dynamic.

To find out what is being said about you in the digital world, Google yourself. Google your company. What is everyone saying about you, positive or negative? Why is that important? Because there are easy-to-implement strategies to counter negative comments. If there is negative feedback about you, that can all change very quickly by building some of your own content, and publishing on your website or social media in formats that rank highly on Google.

Your healthcare partners

Who are your healthcare partners? Who are the doctors, the nurses, the physios, the allied health professionals that you deal with every week? Who do you refer people to? Who refers patients to you? Is there an opportunity to do better partnerships with them in a digital space, but also in an offline space as well? How do you communicate with them? Pharmacies don't traditionally have clinical systems, unlike our allied health and GP partners and specialists. So when we send them communications or patient information, is it in the most appropriate format? There are secure communication examples like Argus that we can use that encrypt the information we send from doctor to pharmacy, and back the other way. Most importantly, it goes directly into their patient record. Is that something of use for your healthcare partners? Would it make things easier for you and them to receive and send referrals?

Your advisers

Do you have a business adviser? Have they advised you on technology in the past? Has their opinion of embracing technology been positive and embracing or has it been pessimistic and dismissive? Do you have an accountant that works on your business? Do you have a bookkeeper? Who is on your advising team? You may have a board of directors that may or may not include all of the partners. It's important to know what they're thinking individually or collectively as they're going to either help or hinder your ability to implement the strategies that you put forward to building your smarter 21st-century pharmacy.

Outsourcing

Do you have any third-party outsourcing? Do you use any contractors for anything? Do you have a team of people in your pharmacy that pack your dose administration aids? Or do you use a third-party packing facility like MPS or APHS? Do their systems make it easier for you? Again there are opportunities in both. You need to consider whether these services will fit in with what you're trying to achieve.

There is a school of thought that you should contract every non-essential service in your pharmacy, especially non-patient-facing services. This could be something for you to consider.

Do you use any outsourcing partners from other countries? Do you use platforms such as Elance or Odesk for graphic design or advertising design? Or even for video editing or creation? This is highly accessible and possible in a 21st-century global workforce. Large corporations have been receiving the benefits of outsourcing for decades; now you can too.

Customer databases

Do you have a customer database, as we've discussed with buying groups and banner groups? Do you own that database? Do you have it in a format that if you decide to do a mail-out newsletter, you're able to plug that data into a service such as Mail Chimp, which can easily and freely manage a small list of customers and have beautiful templates designed and distributed by you or your team?

Sustainability

Do you have a policy on sustainability? Do you use LED lighting, or solar panels? We discussed this earlier in the chapter – the opportunities to be sustainable but also to save significant money are important. The Guild is running a great project at the moment called the Seed Project that can create significant operational cost savings that include items such as LED lighting conversion. I encourage you to look at that.

Membership benefits

Do you have an organisation membership to the PSA or the Pharmacy Guild, and have you looked at the benefits of those memberships to you? Membership of the Pharmacy Guild has a lot of benefits. There are benefits in the areas of workplace relations, payroll, quality care and professional services implementation, both from The Guild and PSA.

Or perhaps they have some IT implications? The self-care program is now available online – you can now email fact cards to your patients instead of having bulk quantities of them just standing in the store gathering dust. You can also have this running on an iPad. Do you use the Guildcare Clinical System to find professional services opportunities? This can be easily achieved by exporting your current or previous month's patient data into a basic Excel spreadsheet. Again, a lot of benefits to consider from where you are right now in building your 21st-century pharmacy. Are these benefits that you're plugging into right now?

IT management

Do you have an IT management plan or, like most people, is that another hat that *you have to* wear? Have you ever considered the costs of having a maintenance agreement, and taking that hat off and giving it to someone who knows IT better that you do? Do you have an expert in-house? How are your passwords managed from a security point of view? You can save huge hours of productivity across a year by using a password management tool.

What is your email system like? Is it accessed by just one person? Do you have a team email address for everyone? Or do your staff have individual email addresses? How do you communicate with your staff currently? A good email system can provide a large number of opportunities for you (for example, efficient communication, mobility, collaboration, etc.).

How do you organise and collaborate with your staff? Do you use a diary? Do you use post-it notes? I guarantee that by the end of this program, you will not.

What are your IT skills like? Is there an opportunity for further education beyond step one?

Your store type

What is your store type? Is it in a shopping centre or a medical centre? These store types have different implications for the future.

If you are in a shopping centre, do you have any lease inclusions? Do you have a marketing fund? Do you have opportunities to collaborate with other tenants via offline and digital means? Do you have an iPhone app? Or does your shopping centre have an iPhone app? Do you have the ability to post offers there? And how do you manage those?

No-one knows your business better than you do

This is a long and comprehensive list of things to consider in the discovery phase, and it is aimed at delivering you a comprehensive overview of your business. No-one knows your business better than you do – you are in an ideal position to know its history, its present, and build the exciting future ahead of you in this program and with your pharmacy.

Follow the steps. It's important that you get all of this baseline information before moving on, because in the next step you are going to need to be able to prioritise. The world of technology has infinite possibilities for you and your pharmacy, but not everything is going to be relevant right now. There is only so much change that a pharmacy can endure in a short space of time. We're planning for 12 months from today. Does it stop after 12 months? No, of course not. But you first need to pick the easiest and most seamless methods of technology partnerships that are going to deliver the highest amount of pharmacy transformation to your business, so it can accelerate towards becoming that pharmacy that you've always dreamed of and that you can't wait to walk into each day.

5

Step 3: Partnership

Well done on completing the second step in discovery. I'm sure you would have found the process quite long. But I'm sure, as you would now agree, you probably know more about your business now than you ever have in your entire business history. What we've put together is a 360-degree view of every process, task and element of your business. Ultimately the reason for doing this – as it has most likely already dawned on you – is that technology coming into your business is not something that you simply plug in and it becomes a separate part of your business. It turns your business into a hybrid of something that you can physically see, touch and handle and something that includes technology to speed processes and parts of it up substantially to give your patients a much better experience. What you have put together so far will be extremely valuable for this step: partnership.

Finding the right partners

Not only do you have a responsibility of partnership with everyone that touches your business, but also to the partners that can help you to implement, plan, train and maintain technology products and services in your business. This is not an easy task, and it should be just as rigorous as selecting team members to join you in your pharmacy. When you select technology partners for your pharmacy, you're aiming to select those who are not just great on the first day they walk in, or great after the first week, or terrific after the first fortnight only to then show a sharp decline in performance. You must demand, as you would of your team, a high level of performance not only in the interview stage, or in the recruiting phase, or on the first day, but on an ongoing basis.

You would also expect your employees to improve and to be able to adapt and evolve in your business, and you should expect no less from your technology partners. You not only need to select the right partners and the right technology for your pharmacy, you also need to ensure that they are both going to be terrific on the first day of implementation and also grow and evolve with your business over time. You need to know that they are not going to let you down, and will maintain a high level of involvement. This is as much as you need to feel confident that your investment (of both time and financial resources) in their services and technology to aid or automate parts of your business meets and exceeds your expectations.

Setting goals and objectives

So part of the second step was to really discover what makes our business tick. What are the important aspects? What are the strengths of our business? And from that, we need to now map out goals and objectives that come from your ideal pharmacy structure from step 2. What I would like you to do is to put together a functional blueprint of how this would work for you, and I've found that a functional blueprint can only come from goals and objectives. As you will come out of this program with a 12-month road map I suggest that you only focus on the top five things that are most important to you. In doing so, I suggest that you adopt some 80/20 thinking – even small changes to the largest segments of your customers can net incredible results for you. So rather than pick on a small aspect of your business that only represents a small amount of your patients, and perhaps also a small amount of financial sales, you may want to pick one of your strengths and really get very good at it.

Your top five priorities

As we know, at the moment pharmacies cannot be everything to everyone, so selecting something that is your strength is of ultimate importance for your successful Transpharmation.

Select your top five goals and objectives. What I'll ask you to do in selecting those five is to do two other things: to set KPIs, things that you can measure. And if you can't think of an element of those processes, just think about what the outcome is. What would be the ultimate outcome? We can work with the KPIs from there. (I'm sure they will come to you in that process.) Rather than thinking of, "I would like to increase my gross profit by 2%," think about ideally what you would like to do in relation to a particular aspect of your business. It could be in a professional service area. It could be in aged care. It could be a compounding aspect if you are already a compounding pharmacy.

I would not suggest using any goals and objectives of any new initiatives. Pick areas that you are good at. It might be in chronic disease or diabetes, for example. What processes and tasks are key to those goals and objectives?

What you should come up with are processes and tasks that are linked to these goals and objectives, and what we are able to then do is to ultimately build a structure that we can then ascend – from bottom to top – by following those processes and tasks to reach these objectives. These are processes and tasks that we already have, from our operations manual particularly. And if you don't have one, then maybe a Quality Care manual will suffice. You can then look at those aspects of those processes and tasks and optimise those by including some technology that aids or automates the process; for example, perhaps in the dispensing process, or utilising electronic forms for your patients to fill out when they first register in your pharmacy, or giving them the opportunity to perhaps register their prescriptions or order their prescriptions via a smartphone app that allows for prescriptions to be sent to your pharmacy and therefore it changes the whole paradigm of your dispensing process (and of course it removes waiting time as well).

Consider all of those technology aspects when you are looking at all of this. I would also encourage you, as the second thing to focus on for each of those goals and objectives, to discover a feedback loop. The feedback loop needs to come from your partners and your team, because it's ultimately important that your goals and objectives are not only set in conjunction with your team and your partners but also that you allow a voice to come from there and that it's not an internal validation of whether you met these goals and objectives.

At the end of the day, we are trying to create patient-centric pharmacies, a pharmacy where patients have their problems solved in the best possible way for them. If we are achieving that then I believe that we are succeeding in what we are doing. Ultimately, as pharmacy owners, have we succeeded in achieving personal and professional freedom? I believe that if we have done those two things then the goals and objectives that we have set are the right ones as far as the long-term outcome is concerned.

Exercise: your top five priorities

Draw up a table with five rows and three columns, or use the one below.

In this table, in the first column, jot down your top five priorities for the next 12 months. It could be your customers. It could be being closer to your partners. It could be how you look at your finances. It might be what's coming up in the future and the need to future-proof your business. It could just be running the pharmacy better by reducing risks and inefficiency. It may be useful to refer to the case studies in the app at this time. But it's important that we set these goals and objectives now as it will be the foundation of your partnership and implementation.

Priority	KPIs	Feedback

Not only do you need to set goals but I also will ask you to set KPIs (key performance indicators). Put at least one or two measurements per goal, so that if you achieved those KPIs you would feel confident that you had achieved that goal or objective.

The third column that needs to be filled out for each goal is a feedback loop. Not only do you need establish and measure KPIs for each goal, you also need a feedback loop so that you can engage with your team, your patients and your partners. Ultimately, if we are all winning together and we are creating a positive experience for patients that everyone can contribute to, we will achieve our goals. And if we can align our team and our partners with every goal and objective we set, we can achieve that a lot more rapidly and a lot more seamlessly.

Getting started

So how are we going to start with partnership? Taking those five goals and objectives and being able to build a structural map of what processes and tasks are included in being able to go from the beginning of the processes to the end of the processes and meeting those goals and objectives over the course of a year, we need to be able to start somewhere and we need to be able to map it. As I mentioned, we need a functional blueprint. Now I encourage you to do this for at least those five goals and objectives, and that may encompass all of your processes and tasks. In fact, it is a good idea to at least map every task and process that you may have, and work out where technology may be able to play its best part in those processes. That way, even if you don't focus on all of them, there are things that you can work on and perhaps even your team can be working on as well, so that when you encounter issues or problems you can revisit those maps of processes and tasks and decide which elements are causing the issue. Generally that's where you will fix that problem. If it's a time issue or an efficiency issue, you may find which technology is a good one to be able to solve that problem for you. You'll have your priority list of what you're going to focus on.

Once you've decided on what you are trying to achieve, and you've got your processes and tasks, you need to start working on what other elements you need to consider when not only selecting the right technology products and services but also who is going to provide those for us as well. You need to be thinking about what budget you have as well. There are elements that your accountant can advise you on, including perhaps what investment you can afford to inject into your business. Bear in mind that the investment is going to be aimed at increasing the level of productivity. The example that probably best demonstrates this for you is a pharmacy automation product – and those potentially can be up to a quarter of a million dollars. When you amortise the leasing cost over five years, this could represent very closely the cost of employing a full-time dispensary technician. Am I encouraging you to change your staffing mix as a result? No. The benefits of re-purposing that dispensary technician – or even a pharmacist in some cases – to improve your business and to grow your business into other professional services is the biggest opportunity, giving your patients a far better experience. You may even be able to reposition that dispensary technician into a triage role, whereby they are involved in the prescription reception process and they can coordinate your pharmacist, whether it be a professional services pharmacist or a general primary care pharmacist. Perhaps this person can coordinate those programs and deal with some of the administration that's involved, as opposed to being involved in a process where they are simply picking boxes off the shelves and labelling and so forth, and leaving that process to the pharmacist who can now engage with the patient on a level that was not possible before.

Now before we go any further, why am I harping on this? Why is this so important? Why has it been so important for me? Well, I'm going to tell you a story. I have always been the biggest believer in trying to improve our business by innovating with technology. One of the most obvious pain points in our pharmacy is the way we manage our stock, or perhaps the lack thereof of management. The majority of us will have point-of-sale systems which have a varying level of functionality, and perhaps we don't even use the functionality to the highest degree. Do we use point-of-sale to forecast orders? Do we use point-of-sale to generate orders? Why not? Is it because we are too slack in stock control procedures? Stock-taking is too hard, possibly. But what we tried to achieve in 2009 and 2010 was a level of stock management that was seemingly unheard of. We were approached by a retail systems provider to pilot a system that hadn't been previously seen in pharmacy. It promised automated replenishment. It promised reduced maintenance levels in managing a database of products, saving staff time in that process. It promised that a centralised database management system would be able to free up a stock controller from a database management role, and the acceptance of orders into the pharmacy would be significantly improved by using radio frequency barcode scanners (which, for those of you who have not seen them, look like a little mobile computer on a hand-held scanner so that you are able to scan the products as you go through). The efficiencies and the benefits to the business were seemingly enormous, and potentially this could have led to further use of electronic shelf labels and automatic checkout machines as well. Quite revolutionary, and we are talking 2009 and 2010.

But where we weren't successful was in our preparation for that partnership, and that forms the basis of a lot of what I will be talking about here – but I shall continue with my story. What we didn't do was a level of due diligence that would've given us ulti-mate comfort that what was being promised could be delivered upon. The pharmacies that were implementing this system on pilot were very small, and we had a very high transaction flow. But nonetheless, we were assured that it could cope with a sig-nificant load of transactions, as it did work in supermarkets too.

The problems begin...

The level of support wasn't properly defined and we believed that we were receiving a ready-made product. But over the course of 18 months, we were essentially the pilot site. We received adequate support and we were able to map each process around implementing that point-of-sale system, and through that we were able to improve the product to a point where we were reasonably comfortable. But that product still didn't deliver the original goals and objectives that were set out at the sale process. We didn't achieve automated stock replenishment. We didn't receive any reduced labour costs or manual costs of maintaining the database, and the efficiency of accepting orders into the pharmacy and invoicing weren't to an adequate level. Ultimately we parted company after 18 months as the frustration was quite mutual that the achievements weren't there.

Why is this so important? Don't undersell due diligence. Don't undersell education. We have already spoken about that in our first step, but it's ongoing, and the education that you have will guide some of the decisions you'll make. You'll know who the big players are. You'll see who's been around. You'll know who's delivering a quality level of service. If they don't, a lot of these suppliers end up with very poor reviews on a number of the review sites, be it Google or some of the technology sites, which I have no doubt that you would've discovered in step 1.

Testing, testing, testing

So the important aspect is – and I mentioned this in the goals and objectives lining up with your team and your partners – that you do small-scale trials, you test with your team, and you perhaps do focus groups with your customers around what it is that they want. It may be bandied as a cutting-edge solution for pharmacy, but is it something that's going to create significant value for your customers' experience? If that answer is no, no matter how cutting edge it is it simply is not important. That will be the ultimate validation for you.

The most important issue: your patients

Once you have mapped out your processes, you then need to map out what the solution may look like for your patients – that's the most important thing. The majority of your priorities will be patient focused. If something helps us to solve patient problems more quickly or enables us to spend more time with our patients, I dare say that will be one of our first goals and objectives that we will be looking at. You need to map solutions from their point of view as well. What are they going to get from it? How is it going to improve their experience? What are the benefits to them? It could be something as simple as providing SMS notifications to patients that prescriptions are ready, or are due to be refilled. How does that work from your patients' perspective? How is it delivered? Is it reliable? Does it work on a scheduled basis? How much effort is involved for the customer?

A lot of solutions in this space will allow patients to interact with you more easily. That can be through SMS technology, or perhaps an app. On your priority list you need to then rank your goals and objectives from the patients' point of view as well.

Your budget

Next, you need to look at what budget you're prepared to put towards these solutions. Ultimately it is possible that some of the solutions that you will implement in your pharmacy will save resources and therefore require fewer staff, or perhaps provide an increase in productivity for your staff so that you are able to contribute more professional services staff – be it pharmacists or dispensary technicians – to other programs that may benefit your patients.

When considering what you are prepared to spend, remember that you may save money in manual labour with some of these solutions. At the same time, you also need to take the view of achieving an improved level of productivity. This may give you the ability to operate a service for your patients that you perhaps weren't able to do with your current workforce model. You need to spend a little bit of time looking at what the likely outcomes are going to be, and how you may want to reinvest some of the savings and productivity of your professional staff into your pharmacy when you implement the changes aligned with your priority objectives.

Pharmacy automation

The example I use for this is pharmacy automation. It is obvious from examining time-and-motion studies that pharmacies will increase productivity by reducing the time professional staff spend – be it dispensary technicians or pharmacists – picking products from the shelf and taking them back to a counter, entering them in, putting the labels on, and so on. It is likely that this process will take less time with the engagement of a pharmacy automation product.

Obviously the price tags of such equipment are quite significant. However, if you introduce such technology you may need one less pharmacist or one less dispensary technician. That may not be true if you are purely picking and packing prescription products, but if your pharmacy needs a point of difference and you need to be able to spend more time with your patients, to give them more value, to solve more of their problems and not just be a supply mechanism, then this is a great catalyst to be able to free up a pharmacist or even a dispensary technician to be able to coordinate some of these services in the pharmacy. You need to look at that, as you may invest in a pharmacy automation product roughly a quarter of a million dollars, and amortised over five years the cost may be similar to an annual dispensary technician wage. Do you automatically remove that dispensary technician from your business? No. Can you re-purpose that person into growing a better and more patient-centric pharmacy? Yes, you can.

It's important that you fully understand the costs and effects of such equipment: it can assist you in building that pharmacy of the future. You must not simply look at the cost in isolation to reduce your operational costs; you must fully explore the benefits it will bring in allowing you to scale and grow.

Board approval and legal issues

For some of the technology products and services, you may need to consider regulatory and legal issues, and even perhaps Pharmacy Board approval as well in your relevant state or territory as some of the products may affect that. For example, if you were going to use display screens in your counselling area, you may want to consider if there is any patient information shown on those screens and could it be viewed by another customer nearby. If not, then it may stack up and it may be allowable. If you were to use tablet computers to be able to engage in the counselling process, again you must ensure privacy isn't impinged upon. If at the same time you were to use technology in how you store the ready-to-go prescriptions, those that have been checked and ready for pickup, you may want to consider again what can be accessed and viewed by a customer.

Some of the legal and regulatory issues are significant, and it depends on the track record as well. Some of the industry solutions will not have a problem. The ones that come from outside our industry you need to consider carefully. So, for example, if you were going to store clinical patient information in a system that perhaps isn't designed for healthcare or clinical use or even pharmacy use, you may need to consider its compliance with Australian Health Information Privacy legislation, and also data storage legislation codes as well. If you don't, it could come and bite you very heavily in the fines that you may have if you encounter a data breach. Ultimately, you need to be confident that they are compliant with your privacy obligations as an Australian health practitioner.

Compatibility testing

Some solutions will require compatibility testing and research. Every one of us has a network of computers and infrastructure in our pharmacies already, and we don't want to disrupt that by adding anything new. We don't want it to destabilise what could otherwise be a very good system. But the system may not be delivering – or be capable of delivering – on these goals and objectives we set out, so we need to introduce some additional functionality via some other technology products.

You may want to consider – and I think this is the best way of looking at this, as we discussed in discovery – looking at your pharmacy as though you were buying it today. Anything that you have in your pharmacy, while you are doing this planning, consider whether you would be retaining it if you were coming into this pharmacy new. Would you run the risk of problems if you were still running Windows XP, which is no longer supported? Would you run the risk of a system or a hard drive which is failing already? Probably not, because ultimately when you weave technology into the fabric of your business (and it is already woven in without a doubt, before we start adding extra functionality), it's essential that it is running with the same fluency that any of your pharmacy operations that are manually processed or carried out actually do. It's incredibly important to do that.

Consider whether you require additional power loading or different data cabling. A lot of hardware products can be implemented without the need for fixed hardware via the use of wireless systems, so that's also important. Understanding the compatibility with your store environment (for example, presence of adequate numbers of power points or data sockets) is crucial, because it's all well and good to put a lot of technology into your business but you need to make sure that it's not coming at the cost of your overall environment.

Sustainability

There are some sustainability issues to consider as well: power loading, emissions and all that kind of thing. They're probably deemed as minor considerations at this stage. It might be LED lighting which can significantly reduce your energy expenditures, and also perhaps even solar panels that may sit on the top of your pharmacy as an alternative energy source. Those types of things need to be considered as well.

On premises or in the cloud?

Any technology for your pharmacy that you consider falls into one of two categories: those things that are run on (your) premises, and those that are run in the cloud. These are two very strong considerations when we are considering our store environment.

On-premises solutions

On-premises solutions that most of us will be very familiar with include our dispensary and our retail systems. They all run on premises. There wouldn't be any pharmacy in Australia that would have any cloud-based systems. It's not to say that they aren't in the development process, but certainly at this stage they are on-premises. So what does that mean for us? It means that we require infrastructure or capability inside our pharmacies that allows for the optimum environment for this software to run. It might be that we need a high-powered server. We need a fixed data network. We need a relatively fast local area network, which is essentially like the internet within our pharmacies. And ultimately that then interfaces with things outside our pharmacy such as PBS Online and EFTPOS systems.

That also means that there is still a significant role for somebody in the pharmacy. Often that would be yourself or perhaps one of your team, who can manage that process when it goes wrong. Someone needs to have a relatively good technology understanding and be able to liaise with software providers to fix issues when they do come up, which is a huge drain on productivity in the pharmacy.

I believe that pharmacies are an environment where we employ people to help solve our patients' problems with us and in support of us. We don't employ people for their administration qualities and we don't employ people for their information technology technical knowledge. That becomes a by-product of inefficiency and inefficient design in our pharmacies. This is an area where we can rectify that. No doubt, on-premises software and hardware requires some capability of team in store, but you can minimise this by engaging with maintenance providers that can remotely log into your system. But generally speaking, they will only act on acute problems. You need to be accessing the support of information technology experts that are able to proactively and preventatively look at your pharmacy network and try to minimise problems before they become functional problems in your pharmacy. That needs to be the basis for on-premises technology, should you choose to continue with any existing solutions or implement new ones.

Cloud solutions

Cloud technology, however, can simplify things substantially for you. A metaphor for explaining on-premises and cloud is that of on-premises representing a private jet and cloud representing a major airline who may operate with the latest technology, like an A380 for example. The private jet (on premises) requires you to own the plane, own the mechanics, and manage the processes of ensuring that the private jet's systems, mechanics and structures are kept finely tuned to provide you with the confidence that not only you and the passengers on that plane are going to be safe but also enjoy the journey of getting from A to B.

Most private jet owners would not be aviation experts I would wager, so it's an incredible burden to take on. You can contract those responsibilities out, as we do with our on-premises systems, but generally speaking because they are not connected to our business strongly they would only act on acute issues that may bob up from time to time.

It's an extremely onerous task as a pharmacy owner to understand the inner workings of a complex IT system on a day-to-day basis and be confident that that system is running at its optimum level day in and day out. Cloud technology can change that. It is like buying a seat on an A380 instead of going through the headache of owning that private jet and being responsible for everything in its complex engine. On an A380 jet from a major airline, you are provided with expectations that the airline will provide you with a comfortable seat, a predictable journey, a predictable experience, and anyone that you choose to take with you on this experience is offered the same. Should anything go wrong, they bear ultimate responsibility and they are charged with proactively and preventatively ensuring that your experience is at the optimum without problems or – worse – failure. It's an incredible amount of confidence that we take for granted every time we step foot on an aircraft. The same applies to implementing complex software and hardware in our business. That's what the cloud can do. It puts the infrastructure into an environment that the supplier has complete control over, and they bear ultimate responsibility as IT professionals and experts for ensuring that the experience is reproducible for you. They will give you what's called a service licence agreement, which will tell you predictably what service you will receive regarding support, training, responsiveness to queries, and also responsiveness to change, to innovation, to education. You will need to have collected and reviewed these documents before you spend a single cent with any of these companies. As it is when we go to a supermarket and buy a can of something, it will do exactly what it says on the tin. That's what we get buying cloud systems. We don't need to worry about putting that into our environment and wondering if it will work, because all the compatibility is given to you up front and the expectations are conveyed to you.

The newer suppliers of cloud software and hardware have what's called APIs (application programming interfaces), which essentially provides a keyhole access to any other services or products that you would like to interface with. Where this is important is when we come back to our mapping. It might be that we would like a particular process to lead to another process or to terminate in another process, or to engage or recruit another process into the best outcome. To do that, you may need more than one program or service. Does any one solution exist where it can cover absolutely every need or requirement of a pharmacy business? No. It is far too onerous a task to be everything to everyone, which is why we've seen software specialists arise in particular areas. And where they can't be specialists there's also niches of integration coming up as well, where there are providers that can take a lot of these cloud software providers and make certain processes from each of them fit with other services and tasks from another. You can design your ultimate workflow around these integration pieces, and the costs are very minimal. You can design a very complicated process map which can be easily implemented. I will repeat, easily implemented *without the need for an IT team.* You can literally decide what software you would like to talk to another at the click of a button.

In our industry, a lot of our core processes are not in the cloud as yet. But that will come soon, so we need to be ready for it. But every other business process can benefit from this right now, so keep that in mind.

Support

We also need to consider what level of support we need and what requirements our store has from these products and services, and also from the suppliers as well. A lot of that rests with the product design and how intuitive it is to our team and ourselves. And as we found engaging with smartphones at such a rapid adoption rate, if an interface is quite simple, such as the iPhone, you can almost achieve any task or process in less than three clicks; such is the design of the phone. User interfaces are now capable of doing very similar things. It makes sense that if the user interface is simple, and we are able to easily find or action what we are intending to do with the minimum of fuss and the minimum of clicks, that our level of support needed will probably go down. The level of IT literacy required is less. That's again something that we can determine by engaging with our team and also with our customers as well. We can afford to be prescriptive in how we want it to work.

One aspect of pharmacy technology that we need to jettison ourselves from is the belief that we must accept what we are being told, that there is no better way or there is no better process or product available. As mentioned, our core products (such as dispense and point-of-sale systems) are largely quite fixed in what they are capable of, but there is choice in that environment. But as for aspects such as accounting or for document management, and other processes and tasks that we are not currently using technology to complete (that is, we are doing them manually), there is so much potential to demand more and for us to believe that we can create a better environment, one that does not become onerous on ourselves and our team and really does start to liberate our time and money. It costs us little or no money to get started at a basic level in most cloud software environments. When operating in the cloud we can have a huge amount of confidence that what we are implementing in our pharmacy does exactly what is needed without error. We don't have server issues. We don't have infrastructure issues. These things can be solved. But it starts with our desire and our belief that there is a better way. And I believe that there absolutely hasn't been a better time to find that than right now.

Maintenance

You also need to consider what maintenance will be required. Even cloud providers will have maintenance, but it won't be done by you or your team. It would just happen in the background, as will updates. Historically, we have done a lot as far as maintenance is concerned. We've had to stay back late to reload databases, and to process upgrades and updates because they existed on premises and that's what was required. Cloud computing solves this problem, and you won't even notice when these events are taking place.

Process automation

Consider the way perhaps you would like to manage your pharmacy accounts. Are you doing a big print run still? Is there the possibility that your software can email those statements to your patients, thus saving you manual processing time and the cost of postage as well. Do you spend time manually generating reports that you or your team need to access? Is it possible that you could simply automate those? Do they arrive in your email box at the click of a button or on a schedule that you choose? It could be a Monday morning.

Can you automate your ordering? Could your system be generating an order for your key categories so that at a particular time of the month your orders are generated for you and you simply need to validate them rather than having to manually build them from scratch? Consider those aspects.

Your team and your in-store implementation

We need to also decide what roles our team will play and what our expectations are of our team in those processes. A lot needs to be considered in this area, and then we need to implement. We need to select. We need to manage. We need to train. We need to maintain these systems. A lot will depend on that, particularly with suppliers and partners you are going to select for that. Are you going to require additional contractors and consultants to help you? Do you have people who have a long history of managing more complex information technology projects? Someone who could help guide you through the process? It could be someone to hold your hand or to manage that process for you.

This is the kind of work that Pharmactive also does. In this program you are getting all the tools that you require, but we are also available to assist you on a greater level should you so choose, which is important to realise and understand, so that you can feel confident in designing this project and managing it through to implementation as well.

You are going to need some tools that will give you ultimate visibility over what systems are being implemented, by who, and in what timeframe. What level of disruption will there be? When will you schedule this? It will be difficult for you to do any on-premises modifications during business hours. Cloud computing can be a great enabler here as well because it can sit separate from everything and only be connected at the appropriate time. That can perhaps be done at a time of your choosing, not of a particular vendor's, because you are able to actually make those connections yourself as well as in conjunction with the supplier.

Project management

You need to look at some management tools for how you are going to manage this project. Just like any big thing, such as planning a wedding or planning a holiday, you need to have a plan for who the parties are, and who is going to be involved. When do they need to be contacted? When will they be coming to the pharmacy? When will they provide quotations to you? When will they provide functionality? What about all the due diligence that you may want to do by contacting case studies and contacting the system providers? Are you getting a consultant to look at how they are going to operate and what security measures their technology business has? Are you able to work out if they want to work with you as well? You need to be able to map all of that out, and you need to be able to say that "today this vendor is coming in", "they are going to install this product", "they are going to spend two hours with our team to train them this week", "they are going to come back the following week", "for ongoing training they are able to be accessed this way as well", and "they're also going to communicate with these technology partners because some of their products overlap". It's important that it integrates well, so that needs to be tested. But, not tested by you – tested by the partners who are involved in it.

Scheduling

You must consider scheduling as well. There is nothing more frustrating as a pharmacy owner to have to disrupt our trade for even the most minor technology change in our operation systems, so it's important that small-scale tests and pilots don't involve live information. This is where it's most critical, because if you are a seven-day business and you are open from 9 to 9, you will have some disruption, no doubt, but you will need to pick the least disruptive time to do these and to perform these implementations. Often with some of the implementations – particularly in cloud software – you just wouldn't know if they had been installed. Generally speaking, the on-premises software that you will use will be the most disruptive to you. Cloud-based software should be less disruptive, because it can be pre-configured and ready to go, and worked on disassociated from your pharmacy work-flow. But for business-critical processes like dispensing and retail, it's important that we really do map out these processes and work out how we are going to do it. If we are implementing some technology that's going to interrupt one of these business-critical processes, we need to consider using off-peak periods to do this. We need to collaborate with our technology partners closely to achieve this.

You require flexibility in your scheduling – it needs to work for you. Some technology partners in the past have only worked nights and weekends, but with the agency of some cloud computing providers you don't necessarily need to conform to those structures. It's important that you make the schedule work for you and have tools that give you visibility over the whole project so that you can see what suppliers are going to be doing for you. When will the supplier be coming into the store to perform tasks? Are they going to be remotely logging into the store? Are they going to be providing staged training on their product? How will that training be delivered?

Who is your ideal partner?

You need to decide what your ideal technology partner, technology products and services deliver. What outcome would you like? In this day and age we are able to set our own criteria and not be told that we *have to* use this or that product. There are new and innovative products in almost every aspect of business being developed every day, so the ability for us to find a product that suits us and also to inject some customisation into those products has never been easier. Don't accept limitations. You can always achieve the result that you're after, though sometimes that could come at a very high cost. Some of the newer, more adaptable technology partners are very flexible and they offer non-prescriptive software that you can tailor almost to your exact specifications, which is incredible to consider. You need to do some assessment and some scoring.

That also may include – as we've already mentioned – levels of integration. Where a process may involve more than one piece of software or hardware, will they talk to each other? Is that possible? And again, I dare say that anything is possible, but some integration is easier than others. It depends largely on whether it is on-premises or cloud computing software. My experience is that cloud computing is much easier to integrate than on-premises, but on-premises integration is not impossible.

Collaboration and selecting the right technology partners

What collaboration will be offered? What level of consulting do these technology partners have with you? Will they help you determine which products will suit your needs? Or will that be a self-guided process you may have to do on your own or with a trusted adviser? Do you already know these suppliers or do you need an introduction? Sometimes these introductions, as we have determined in our first chapter, can occur at a trade show, via video, or any other communication process.

Who is the most knowledgeable person in these companies to be talking to for your particular purpose? And how accessible are they? Because, again, nothing is worse than dealing with technology partners that are largely inaccessible, particularly when things go wrong. You may be someone who prefers phone support or face-to-face support if need be, but if that supplier doesn't offer it, that's something you may consider when deciding whether you want to use that partner as well. Obviously you need to be doing some comparisons between suppliers.

Nothing is more frustrating than having five – or maybe even ten – technology partners that don't collaborate, so that you find yourself having the same conversations over and over again. You need a good method of communication between technology partners and how they are going to tell you what they are up to, and you can tell them what you're up to. There are some tools that can help you with this too.

Then there is also the period of time – and it may come a little bit earlier – where you need to introduce or just even start the discussion with some vendors and partners. How do we go about doing that? Sometimes that can be achieved in the form of a conference or an exhibition, where you can have introductions to a lot of this information. Certainly there are many different areas – as we now know from step one – that you can understand what some of these partners or suppliers are capable of doing. But at the same time, if you are going to use a project manager to bring this all together, that's something that they can perhaps do on your behalf as well and they can take the functional scope of what you have designed and go and find the right people who are going to fit into that framework, which is why I underlined again that the functional blueprint of what you would like is most important, and for you not to hold back on any aspect that you would like. I believe that the technology is there to be implemented in whatever function you want to achieve, so it's important that you don't undersell that process, or you will be disappointed in the overall result.

You then need to be able to compare the vendors on a side-by-side basis. You need to decide what the most important aspects of each technology partner are in the overall process of getting a particular product or service into your business. Is it reliability? Is it backup? Is it security? Is it privacy? You need to be able to determine on a level playing field what all these partners are capable of. Do they innovate? How often do they innovate? How do they do updates? Are they on-premises? Are they in the cloud? Do they integrate well with other systems? Or is that something that they provide the tools to be able to implement? You may now know what an API is. It allows you to have a particular piece of software plug into another piece of software quite well. Will they provide one of those, and whose responsibility is it to integrate with other systems? Have they done that proactively? Are they capable of plugging into an existing integration directory of products?

All of these things need to be considered. And it's not just a case of considering capability, but we'll come back to our functional blueprint and be able to see whether some of those integrations are important or not so important for what we want to achieve. Based on that, you will give them a score and then decide on balance which one you would like to deal with.

Ultimately the most important aspect of any technology that you implement is that it functions well for you, not only on day one but well into the future. You may have technology partners that have a great product but have very poor support. As you know through this process now, education is key, not only for today, tomorrow, next week, or next month, but well into the future. And that relationship is an ongoing one. So don't select technology partners that won't have an ongoing relationship with you.

Data management

In considering these partners, you're also looking at the data management and the maintenance as well.

Where does your database sit? Where will a technology partner keep data? You'll find this information in their service licence agreement. They will need to demonstrate to you what level of redundancy they have as well – if one of their data centres was to go down, what backup do they have? How do they justify the level of up time where you don't miss out on any functionality because the software has failed because it can't be reached? What redundancy strategies are in place for that? Down time might have something to do with their end.

Some point-of-sale vendors require you to do database updates, every day, every week, every month, or on a scheduled basis. Is this perhaps a service that one of the technology partners can offer so that you don't have to employ someone to do that, in which case you are employing a technical person to do that for you?

Does the data work well with other programs? We used the accounting example earlier. Is the capability available with the existing system to be able to pass from one system to another without it needing to be double-entered? Another very good example for this is in medication management systems. There is currently no automated way of having medication information transported from a dispense system to a medication management system. There are several that allow you to have an interface where you can drag information from one system to another, but none that does it automatically. That's an additional process that you need to include. There are some lower functionality medication management software systems that allow you to do that, but perhaps their clinical aspects of the clinical software aren't as good.

Again, it's a balanced environment and you need to consider the time taken to get all of these things working well for you.

Your internet connection

You also need to consider that the pathway between the technology partner and you – when dealing with cloud computing – is the internet. One aspect that you need to consider is that cloud computing is best enabled by having the best possible internet connection. I emphasise that for you. If you use on-premises software, you remove that as a dependency. You will be able to function because you are in control of the communication because it is all local, and it doesn't require information to leave the pharmacy and then come back. But as we've discussed, the complexities of maintaining that is often why it is not a sustainable model for a pharmacy let alone any business, particularly small businesses that can't afford IT departments on site.

The best possible internet connection might be cable, ADSL2+, or soon the NBN. But what happens if we lose our connection? How will our pharmacies be run? Our patients don't stop coming into our pharmacy because the internet's down – that shouldn't bother them. So do we have our own redundancy, and are we counselled by our IT partners to have backup systems in place, such as having mobile broadband devices that we can simply plug into our core systems to allow that process to continue? Perhaps more critically, does the software – and even the devices – enable an offline mode as well, so that you may be able to continue to dispense and to continue to process retail transactions? Is your database kept locally as well as in the cloud? As soon as your internet connection is restored, your data should synchronise back up with the cloud and you're perfect again. Your patients wouldn't even notice the difference.

Payment systems

Another consideration is payments. We sometimes have a delay of service from a particular bank. We have always opted to maintain a second electronic funds transfer line. We have an integrated EFTPOS system which runs on the internet, but we also have a fixed line which runs on the telephone line. You can also have mobile payment devices which the banks can provide, as well as those that interface with smartphones, which involves having a card scanner attached to your mobile device. Considering this in your overall strategy is important to make sure that your patients are not inconvenienced because of any connection issues.

Interoperability

We need to also be looking at the data that we are keeping to see if it is what's called *interoperable*. Is it able to communicate or be passed through via integration to another system? Often data systems are designed particularly around the database and also the common elements that can be tagged and linked together between separate systems. That means your processes can either be aided or automated by the various systems of technology, and you can achieve the highest level of efficiency possible.

Mobility

We need to also consider the products and the services and how available and accessible they are in a mobile environment, with the need for us to deliver services to our patients not only in the pharmacy but perhaps also in their homes. That's going to be no more evident than in the upcoming years when our ageing population will be living longer. Our aged-care residential facilities will not be able to cope with the requirement for the number of beds, so our older patients will be staying in their homes longer. We will need to be able to come to them, so mobility is of ultimate importance when you are considering these systems.

In a patient-centric pharmacy, we need to be where our patients want us to be, and that may be visiting them at their homes. We already do that successfully with home medication reviews. There is some great software available to enable you to run a very efficient home medication review from a tablet computer as opposed to needing paper of an offline laptop. You can go into these interviews and walk away with all the information already inside a template that you can then turn into a report, as opposed to it needing to be handwritten or even transcribed, or even transported from one system to another, like a Word document into another system.

You need to consider the ability for mobility. It might be health information. You might want to come into the counselling area and be able to show a patient on a tablet computer a patient education video. It might be collecting patient information or customer information for a loyalty card. You might have patients filling out their forms on tablets as well as opposed to having paper-based solutions. All of these things you need to consider if you are going to decide if a technology partner is the right one for you.

Security

What level of user security is provided to prevent team members from accessing parts of a particular program that they shouldn't?

How susceptible is the system to threats? We have seen in 2014 the heartbleed bug, which exposed thousands of user names and passwords, and many cloud-based services were affected by that. But they were able to put a stop to this very quickly, and were able to assist their users by telling them if they were susceptible and whether it was appropriate for them to change their user name and password. So, largely, they did extremely well to be able to allay the fears that our user names and passwords had fallen into the wrong hands.

Location and installation

Location and installation are important too. Having cloud computing allows us to have flexibility of where we want things installed as far as hardware is concerned. The majority of such systems can operate on tablet devices, but at the same time we may want the structure of having permanent fixtures. It might be stands, or brackets that may assist us in having a permanent structure involved. Where will that occur? It will then need to be designed around the physical workflow inside the business as well.

For on-premises installation, you need to also consider where you may want to set up your infrastructure, such as in an area that may have very good air circulation and cooling, and also access to what's called *uninterrupted power sources*. Should you have a blackout or a delay of electricity access, you are able to safely shut those systems down either manually or automatically while you wait for a return of service. Those things need to be considered too.

Updates, innovation and feedback

Updates, innovation, and also the customer feedback loop from the product suppliers and the technology partners are incredibly important too. You want to be able to feel confident that bugs that occur get fixed very quickly, and that patches or updates don't take weeks or even days – they can be fixed very easily. In the past, many of our traditional pharmacy vendors have taken long periods of time to fix bugs and to change things in their software systems, and often those things needed to be run over-night to ensure that they were properly implemented. The newer technology partners are capable of doing this almost immediately without you even knowing that there's been a problem. For example, any bugs that you may have seen in applications you may have downloaded in the smartphone can be updated in almost the same day or hour following the presentation of that bug, such is the speed in improvement of updates. But also you want to know what these partners are doing. What's their big game? What's their structure? What's their history? Do they want to be a 100-year company? That can give you the confidence to deal with them as well. Or are they part of a bigger company and they're just fulfilling a small need hoping to be absorbed into a bigger company? These things need to be considered too.

We also need to understand what their competitors are doing. If we select one particular product, we need to also be aware of what the competitor products are and be up to date about what's going on as far as innovation across the competitors, should there be the need to benchmark those and optimise that, which we will cover in the next chapter.

Charging models

What's their charging model? Traditional software in pharmacy has always been a monthly fee. But some of the newer partners may give you the option of a discount for an upfront payment, and they may also have a tiered level of access. We're increasingly seeing software providers provide a limited version of their product for free or a limited cost. Then your level of use will determine what you need to pay for. Typically, a lot of these programs that solve particular problems for small businesses are designed to scale, so that as your business grows you also require a higher version of their product. I guess that makes sense: if their tools have enabled you to grow, it's natural enough for you to require more scale from those tools to be able to deal with a higher number of customers. The best example may be an email marketing program that you may want to use. You may only have a small group of customers that you want to target initially. For sending fewer than five hundred emails, there is probably enough functionality for you to get what you want at no cost. Then, as your customer numbers grow, you are probably going to need to upgrade to a higher level plan. That's just an example of where a charging model may be useful.

System requirements

What are the minimum system requirements and power requirements? What are the hardware requirements? Do you have tablet computers if we are talking cloud providers? Do you have internet access? What level of internet access have you got? These things need to be considered when you are going to be using those. There's no point implementing an application or a product or service that places huge demands on your business that it's not capable of delivering, things such as an application that may require the use of a tablet or a smartphone and you don't have the capability of charging it on a regular basis. It might be an application that you send out with your delivery driver which requires a high use of GPS, therefore it drains the battery significantly. You must have charging capability or a special case available for those team members – to ensure that they don't have any loss of service. All those things get factored in.

IT literacy

With the products, what are the IT literacy requirements? Is it intuitive? Does it have an easy interface? Perhaps the best and easiest interface we have seen is the iPhone and the iOS platform, in being able to find things on an iPhone in roughly one to three clicks. If you were going to use, for example, a cloud-based accounting system, or any accounting system for that matter, are you going to be able to do it with a minimum of fuss? Is the interface something that's friendly to you, because if it's not then it's not worth the investment in a product that you are not going to be able to easily use. A lot of these user interfaces are becoming more humanised and less technical, so that you are able to do fairly complicated tasks in a very simple manner, because of how it's displayed to you and the requirements of the software.

Partner location

Are the technology partners local, national or global? Obviously, the bigger the technology partner, often the more reliable. But also the support level may be different as well. A large proportion of these technology partners are implementing self-service capabilities, and you are often able to find answers that you need without needing to talk to a support system directly. These things all need to be considered when we are looking at the right technology partners.

What will the support be? Are you comfortable with self-service? Are you happy to receive a step-by-step map of a process and work your way through? Or would you prefer someone to actually step you through this, and also deal with any conflict that you may have when dealing with that process? These are also some things needing to be considered. Often the smaller vendors may be able to give you both, but often the more personal support will cost more.

Are they accessible? Are they flexible to Australian environments? If you are dealing with an American company, perhaps such as Google or Facebook, the need to be able to contact them directly may not be that great. But if you do need to, what would they provide for you? Probably in the case of Facebook, not much. But in the case of Google, the more complicated the product and the higher the cost of the product that you are purchasing, usually the higher the level of support that is required.

Often the best way of engaging with the technology partner and knowing what they are going to do for you is reviewing the SLA or service licence agreement. That will detail, when you sign up with them, what they will provide for you and also in what timeframe. And also in terms of operation, what timeframe they commit to for maintaining their services. The service guarantee or 'up-time' percentage is usually between 99.5 and 100.

Trials

Will they provide you with pilot testing or a trial environment, or a 30-day trial before you enter into an agreement with them? If you've set up a dynamic feedback loop, you can get rapid feedback from your team, your customers and your partners as to whether the trial product or service is giving them or your business an improved experience.

Training and on-boarding

What method of training does your team respond best to? Do they respond best to face-to-face training? Or if you have part-time staff or casual staff that attend your business infrequently, would they benefit from a flexible approach where there is online delivery of training? The ability to hold their hand for the first few times using a particular process can also be beneficial.

Do they have videos? Do they have interaction capability that allows you to better understand their product or service, that enables people to get on board quicker without the need to wait for a trainer to attend the business? Is there an ongoing cre-dentialing process beyond your first implementation? Do you require some credentialing or certification to continue using a product? If you have to bring on someone new, what do they need to do to get up and running? Do you need to send them away from the store for a day or two training? We would hope that time away from the store would be minimised – we cannot afford that sometimes. Flexible learning around those things will be important, and also important around the scheduling and design of your project.

Training is important, not just for you in your education phase but for your team. Is it flexible? You may have full-time key staff members who you want to get the best possible training face-to-face, but that may not be possible for you and your team in its entirety. Remember, you may have part-time team members that work elsewhere or they may be uni students, and they need to be able to get the same level of understanding that you or a full-time staff member may have. Otherwise, it is your patients who will suffer because of their lack of training and education around the technology that you've used to improve your patient care. It's important that technology partners provide you with a flexible learning environment, and the ability to train your internal staff.

When you do agree to take on one of these technology partners, what is the on-boarding process? Is it something that all your team is going to use? How easy is it to get everyone into using that particular product or service? What training is involved? What commitment per person is involved? Do you need to have different levels of training for different levels of IT literacy? Possibly. Is there a train the trainer type of environment where they can train a couple of people in your business and you appoint a store or group champion of a particular product or service that understands the product/service on a very high level and can teach others?

The days of having trainers come to your business and having to go to a training centre – I believe – are very limited. We don't have the time, and we can't afford to send our staff to training courses for long periods of time. We just need them to be become more adaptive. But at the same time, the product quality needs to be improved so that it's not so difficult to actually start to implement a certain process. That sometimes comes down to the training information that they have available as well.

And – as we've discussed – what about additional new training systems? Is there a certification course available? Is there a specific organisation that they can put you in touch with that deals with high levels of training and education requirements that they may not offer, so that you are able to get the level of training and implementation service that you may need?

Industry knowledge

Do your technology partners have industry knowledge? Some of these global partners that we are talking about will have absolutely no understanding of the Australian pharmacy industry. To understand a system's impact on your industry, you may also need to access a trusted adviser. They could also work with your technology partners to ensure that they understand your obligations as a pharmacy owner in Australia. Perhaps they can also – even from a global perspective – provide you with case studies and white papers to demonstrate its use in a health care environment or a comparable industry such as perhaps hospitality, where you can see the functionality work and validate your decision.

Exiting the partnership

Of course, before you enter into a partnership with these technology partners, consider also your exit strategy. Some programs and services may only be useful for a certain phase in your business. If you were to grow or to add additional services or even additional locations to your pharmacy as you build an empire perhaps, your requirements may change. So if you needed to get your data from these services, how easy would it be? Do they allow you, even as you are going along, to keep a local or an external backup from their systems? There are now independent vendors that will back up your database away from the cloud partner's infrastructure so that even if they fail to deliver on their service licence agreement and their redundancy strategy fails, you also have an additional backup to fall back on as well which is quite important.

The new world of business

Now you may be thinking, "This is an onerous task to take on." But as with managing general business resources like finance and people, the more you know about your capabilities the greater your chance of maximising them for the benefit of your business. We are in a new world of business that can be rapidly improved for our patients' benefit if we engage the efficiency and best practice processes that technology can bring. We can also afford to dream of what can be the ultimate solution that we can provide for our patients, and that can carry across many different areas. I believe that the pharmacy of the future must be a hybrid of physical and digital. Our patients aren't always going to be finding us by walking through our doors. We need to not be bounded by our location because our patients may not want us to be where we are right now. They may choose to go on holiday – but as their pharmacy, we can't go on holidays with them and be as accessible for providing advice when they're away as we were at home. If they want general advice, or they want to order products from us, they should be able to do that from anywhere they choose. Life sometimes is not predictable. Our opening hours cannot match their requirements on all occasions. We've seen pharmacies in Australia try to operate 24 hours a day, seven days a week. It isn't sustainable – but in a digital environment, it most certainly is.

We need to consider where we fit in this new world, and also consider what our expectations are of other retailers that we deal with: our supermarkets, our banks, our travel agents. What right did we have to tell our travel agents 15 years ago that I want to be able to book a holiday at 1 o'clock in the morning? What do you think they would've told us? "No. I am open 9 til 5. You can come and see me in my office Monday to Friday." Are we bounded by that location and time restriction now? No. The same with banks. You had to be in the branch within certain hours to withdraw your money. Then ATMs came along. Now, since the world can work electronically, we don't even need to access our money in a physical form on most occasions because it can be handled electronically. When you consider the transformation that was evident in these industries, our industry is next. Which is why we have Transpharmation to get us there. There's never been a better time for you to implement this and ensure that your customers are able to experience this from you and not from your competitors or a bigger chain.

Partnership – as we discussed earlier – is the most important step because it ultimately underpins the foundation of a successful plan. If you have stable, effective partners working with you, and they're aligned to your goals and objectives, then your probability of success in implementing this is vastly improved. The implementation plan that we are putting together here is designed for 12 months, but no doubt the considerations that you made here will be relevant to any future decisions that you make and anything that you may want to implement in your pharmacy beyond the next 12 months.

6

Step 4: Optimisation

As we can appreciate from partnership, we've now got our five key areas. Now you might be asking yourself the question, putting this framework together and working it out, *how am I going to do that?* The accompanying app will guide you through that process, and includes detailed case studies covering how you can implement these changes in your business, using my experience to guide you.

What can you expect in these case studies? Every aspect of what we have spoken about so far is covered: education, the discovery using a real-life pharmacy case study, and also the partnership and implementation to make that happen. If you examine these carefully, you will be able to use them as templates to help you put everything together.

Why we need step four

Now you might wonder: once partnership is done, once the implementation is done, why is there a fourth step in the process? Surely we would be exceeding our original expectations because we have partnered with the right people and they are helping us achieve the results we want at this point in time. The fourth step of optimisation is the one that turns a good business into a great business – that is not only great now, but is growing and is sustainable, and it will continue to be great long into the future.

As we can now appreciate, the business landscape has changed forever. The technology landscape has changed forever. You may be wondering why I don't mention many specific brands and companies in this book: it is because they change and evolve every day. You must always be reviewing and measuring the success of your current business and – particularly in this 12-month program – looking at your goals and objectives, and looking on your KPIs frequently. You need to be monitoring them and reviewing them so that you can feel confident that you are on track and that everything is functioning as expected. That's the key to the success of Transpharmation. The first three steps will get you to what I would consider the base level of understanding and knowledge. In addition, it will help you find the key priorities in your business that you may want to target and fix and optimise initially. But that's not maximising your business potential and your future plans, and also acting on the ideas/insights that come from the strong feedback loop that you set up between you, your partners, your team and your customers. That network and feedback loop will allow you to optimise effectively.

The importance of feedback

So it's no surprise that the first step in optimisation is reviewing our goals and objectives, setting new ones, altering existing ones, and monitoring our KPIs and our feedback loop, and doing that often. To do that effectively, you need to have a strong feedback loop set up. It needs to be set up so that you can react each week or each day, and there's constant access to you and your team to be able to provide feedback and you are able to action it and respond quickly and swiftly, as opposed to the age-old suggestion box that may only get emptied once a month – if at all. This can be done via any method of communication that your network is comfortable with (in-person, online collaboration, surveys, email, etc.).

It's important in this process that we are dynamic and that we are able to validate our decisions very quickly overlaying our goals and objectives, taking on board the 80/20 thinking discussed in the previous chapter, we need to continually focus on the strongest aspects of our business. That will enable us to ensure that the majority of our customers become early adopters, and also engage in these optimised pathways that we are developing for them. This includes the technology that is built around those strong processes and practices that underpin our businesses from the very beginning. Ensure that you *validate* everything you do or plan to do.

Negative feedback

In doing this, you may want to repeat your pilots and trials. If they don't work and you receive negative feedback, that's sometimes a good thing because it means that perhaps you engaged some early adopters, but you haven't quite got it right yet. You may want to change it a little, get some more advice, speak to your technology partners, and find the right fit for your customers and your patients, remembering that if they're not getting the benefit then your validation isn't there and you need to continue to work on your implementation. You may want to repeat some of these steps. The key here is to fail quickly; this will ensure that lingering frustration doesn't manifest itself in the feedback loop, as failing aspects or functions are quickly removed or improved to acceptable standards.

You will continue to learn. We can rediscover and remap our processes so that we are able to put together another small-scale test or a trial that a group of our customers can evaluate and – hopefully – validate that we are on the right track in what we are trying to achieve.

Involving your customers

Let's look at an example that we used in the previous chapter: customer accounts.

How will your customers respond to electronic sending customer accounts and statements? How do you find out? Take a group of 12 customers and trial it. Take some customers with less technological ability and some who are heavily engaged with technology. If you run into problems along the way, you need to consider how you are going to overcome them. Do most of your customers have email? If they don't, it may not be a good strategy for you.

Another aspect might be sending email messages out to your database. Again, if the majority of your customers aren't on email, it may not be a good fit. But if you have a 50/50 split, you may find this is an effective strategy. You can email customers who have email, and print out the exact same information and have it available in the pharmacy, also sending it by direct mail. If you have a database of addresses you can customise those communications by name and address. These things can form part of your local area marketing strategy.

Customers can sometimes be scared of change, but generally this occurs when they are not informed and they are not part of the process. The way to avoid this is to approach a group of customers and say, "We would like to improve your experience. We are thinking of implementing this process and using this piece of technology. Would you be willing to try it?" You can also get feedback from them during the process.

It could be something as large as implementing a pharmacy automation system. You may want to talk with your customers and say, "What would you think if we were able to reduce your waiting time and you are able to spend more time with your pharmacist in consultation? What if we were able to do additional things for you like measure your blood pressure every time you come in, perform diabetes checks, review your diabetic blood glucose readings, go through a process of looking at patterns of those glucose readings, and also confer, communicate and collaborate with your other health professionals? Would that be of interest to you?" By implementing that one technology, you enable stronger patient engagement and you will be delivering so much more value to your customers than just simply removing the waiting time. It can be much, much bigger. You need to think big about what you can achieve.

Technology partner optimisation

Optimisation may come up where your technology partners come to you with a new innovation or update. That update may include increased functionality that could improve or optimise some of your existing processes by enabling you to streamline or automate more of that process. You may want to become an early adopter of that and roll that into your strategy. But again, taking on board what we've learned in step three, we need to ensure that we get regular feedback about anything that we have changed, from our team, from our customers and from our partners who are affected. If we do that, we are less likely to create road blocks where customers do not want to deal with us anymore because we keep changing and appear unstable.

Customisation

We may be able to go further with some of our technology partners and customise parts of their software to enable greater automation in our processes. It could be that we are able to brand some of the products that we utilise so that our customers see it as coming from us. That may be, for example, when you are using an email marketing program; you may want to have your logo and your details included as opposed to just the product brand.

STEP 4: OPTIMISATION

It also may be that from a customer relationship management perspective you might want to collect customised information about your customers. It could be information about their health conditions, so you can segment your database by a particular health condition. That can also be maintained by having a customer-facing portal where customers can fill out a digital registration form when they first come into your pharmacy, and you enable them to update this so you have a permanent record of the important information in the pharmacy. This can be utilised to improve the business's activities by noting the segments of your database that may have a particular medical condition, and also being able to generate custom reports. You may want to know who in your database is a diabetic. It can be very difficult to do that in existing point-of-sale systems. Customer relationship management systems enable us to do so much more. We can often customise the information we want to find out about our customers, but also to give us more knowledge to be able to go further with some of our customers to solve some of their deeper problems. With the right systems we can do that proactively instead of on an ad-hoc basis, which is probably where we are at the moment.

Automation

We have looked at increasing the levels of automation in our pharmacies. It could well be that initially some of your processes and tasks involve manual data entry or other manual tasks performed by certain team members, and you may be able to automate that.

The seven-year document loop (case example)

A great example of optimisation is the management of financial documents in your business. Is there a way to minimise the number of times a piece of paper is passed through the pharmacy team, your administration through your bookkeeper and accountant, and also archiving? There absolutely is. Let's start with the documentation that we encounter when we're ordering products in our pharmacy. One of your team may on an ad-hoc basis or a daily basis order products into your pharmacy via your wholesaler, or a direct supplier. A purchase order needs to be generated, and that can be done from your point-of-sale system. It can also be done via a phone order, a fax order, or even a representative coming into your pharmacy and proposing an order that you may then send on to your supplier. Or they may do that on your behalf. But nonetheless, there should be a record in your pharmacy of that order being placed, so that your stock or inventory manager can monitor the order process as it goes through your supplier. You then have the ability – if you are using your point-of-sale system – to transmit that order, and get a notification whether that order has been transmitted successfully. You may also be able to view this via a customer-facing portal, where you are able to view all outstanding orders you have placed with that supplier. Most wholesalers and direct suppliers offer this now, and you are also able to update your details there. You can get a delivery notification slip that you can print off if you choose. You may also then receive a notification from that supplier once the order has been dispatched, which may be directly from them or from a logistics partner. You may then also receive an electronic invoice, as the consignment may not include a printed invoice to save your supplier money in paper and printing. You may need to print that out. So, your stock or your inventory manager may need to have the purchase order, the invoice, and then you may receive also a receipt of delivery from the logistics partner or an internal courier from the particular supplier. That may include a picking slip that you are also able to reconcile with the other two documents: the purchase order and the invoice. That's already a lot of paperwork!

Your staff may then use an electronic device, such as a radio frequency scanner, or a manual method of ticking off the individual invoice to ensure that all products ordered can be reconciled against the purchase order and the invoice. They can also note any discrepancies, so that if the wrong products have been delivered – by either user error or by supplier error – that can be brought to the attention of the supplier at the earliest possible point, so you can apply for a credit or a return authorisation… generating yet another document in the process.

From this point, the stock can be put away in the store and your stock levels within your point-of-sale system can be updated by entering that invoice into your point-of-sale system, either automatically – if you are using a radio frequency scanner – or manually. You may also at that point note any price discrepancies, whether that be a reduction or increase in cost price. You will then need to update your margins. From that point, you may also need to generate shelf labels if there are new products or new prices that need to be applied, and the invoice will be accepted into the system.

The invoice generally will then be put aside, and entered by the same person from scratch again into your accounting system, ensuring that all GST codes have been accounted for, special accounting codes have been attributed to that order, and that the total of that order matches what you have received from that supplier. Obviously if there are discrepancies, such as being short sent or applying for credits, these issues need to be flagged so that your accounting entries in your general ledger are 100% correct when compared to your point-of-sale.

The majority of us don't have on-site bookkeepers, or even a trained person in the pharmacy business, to be able to manage all of those processes. It is quite labour intensive and highly administrative and not very patient focused. In these tough times, we can only afford to employ staff who are able to engage with our customers, so we may look to outsource our bookkeep-ing. We currently may have an accountant do everything for us, including the entries. But they don't live inside your four walls of the pharmacy. So then that document needs to be shipped off in paper form to that accountant or bookkeeper for it to be re-entered. This can cause significant delays. Once it is entered, it will then go into an archive box which – for the majority of invoices and purchase orders – we need to keep for seven years. This is a very long and convoluted process – as you can see! The documents move many different ways and stay in circulation for a long period of time.

Now the document doesn't stop there! After seven years, it needs to be securely destroyed, so someone needs to be able to recognise when that seven years is due and make the necessary arrangements for that document to be securely destroyed. Only then have we completed that document workflow journey, seven years later.

STEP 4: OPTIMISATION

There must be a better way. Read on to find out…

Optimised document management

So you may consider how you are going to optimise that process. We can utilise tools such as electronic document management tools, not those that can just store them but that can also dynamically extract information as well. There are document management tools available where once you have received these documents, you can either scan them in at your pharmacy or send them away in specially generated envelopes to be scanned on your behalf, and have the data extracted and placed in a secure online portal which is compliant with all Australian Tax Office (ATO) regulations and storage and privacy requirements for an indefinite amount of time. Thus that document becomes a collaborative document that can be accessed by not only you but your team, your bookkeeper and your accountant for as long as necessary – but it doesn't stop there. That can also interface with a cloud-based accounting system.

For those who aren't using cloud-based accounting systems yet, this is where the real efficiency comes in. In the past, accounting systems usually only existed on premises, and this meant that there was only one database. So, if someone else needed to work on it who wasn't in your business, you would have to create a duplicate or a backup and physically hand it to them. Then you couldn't use that database until they were done with it and you were able to merge the data back into your program. Very inefficient. A cloud-based accounting system enables everyone to work from the same database. The supplier will back up your data in multiple locations and have a data redundancy strategy in place to ensure that there is no loss. They also offer mobility access via smartphone or tablet, so that you can feel confident you have access to your finances anywhere, anytime you choose. It's incredibly powerful. The time that is then taken from invoice acceptance in your pharmacy to the end of the administration process is ultimately one or two days (when your workflow is highly efficient) – instead of seven years!

Why only a few days? Well, once the document is stored electronically in an ATOcompliant repository, you no longer have to keep that physical document – the document management provider can shred it for you on your behalf. Thus, in seven years' time when that document no longer has relevance to your business because it no longer needs to be retained, it can either be deleted from the database or it can be retained as cloud-based storage has an infinite storage capability, which is incredibly powerful. Now, we no longer need to use up our valuable pharmacy space to retain archives, or spend additional money on additional storage space, and spend additional money on secure document destruction for these documents. There's a huge amount of time and money that we can save from this process, not to mention the confidence of knowing that this has been automated for a fixed monthly price which is potentially many, many multiples less than what we are currently paying.

This is a very good example of how technology can streamline, organise and optimise what is a very complex administrative procedure.

Revisiting goals and objectives

As part of the optimisation process, we need to regularly revisit our goals and objectives. Optimisation must be done on an ongoing basis. It's something that you need to be reviewing every month. Even if you've just put in brand new systems, programs and services, you need to be reviewing these every month to ensure that your goals and objectives are being met. That can be flagged by the KPIs and your feedback loop. You may even want to pull everyone together for a workshop once a quarter, to really pull it apart and decide on future steps for how you can improve the process. How have your customers and your team responded to the changes?

Continuous quality improvement

Continuous quality improvement underpins optimisation. We just need to continually get better. Pharmacies for too long have sat in the same business model and have had the same customer delivery model. We need to be creating together with our customers. We need to be engaging in co-creation of what our business and our pharmacy can do for our patients. It's no longer a pharmacy owner's role to take 100% responsibility for that. Our patients now have the ability to create that with us, and they will if we engage them. This may highlight problems that ultimately improves essential services for them. Because we have mapped our systems, when our customers cite areas of concern we can now easily identify the areas in the system or workflow needing improvement.

Outsourcing

Another way in which we can improve processes and improve our patient engagement with our current staff is by outsourcing. We mentioned a little bit earlier how we can outsource book-keeping and even accounting from being an on-premises activity, and you're probably quite familiar with that. But there are other capabilities that once we have our processes mapped out and we are comfortable in how they can be completed – particularly our administrative functions – that we're able to outsource to anyone.

Outsourcing overseas

It doesn't even need to be a team member who is employed in this country, it could be to the most appropriate and highly skilled person to be able to engage in that process, who could be in another country! Using the world's economy as leverage, we're able to access a workforce that may have lower workplace remuneration expectations than what's currently available to us in Australia. Many huge corporations in our country are already doing this for administrative functions, and also for marketing and communication functions as well. The key to success here is mapping your processes and utilising cloud technology that enables you to closely collaborate with your team either at home or abroad.

To have on-premises processes outsourced to virtual team members outside of our country requires significant investment in remote management software to enable access from that end. It's not impossible, but it is a lot easier if it can be done with cloud computing models. Through the agency of tools that manage passwords, for example, you can feel confident that you're engaging virtual team members who aren't simply going to take your valuable information and run off with it. There's obviously a lot of processes and guidelines to finding the best virtual team. This comes under our umbrella of optimisation, because, as you get more confident with these processes and the use of technology, you can hand over some of the management to a virtual team member, as opposed to either yourself or a current team member you have in the pharmacy who you could re-purpose to improve their productivity in improving your patient care.

Outsourcing within Australia

If you are uncomfortable with outsourcing away from Australia, there are virtual offices within Australia, bearing in mind that the costs do go up. One thing to be aware of when you are considering the cost of virtual team members is that it is always related to the average income in that particular country. To employ someone in the Philippines for a set dollar rate per hour may seem like under the award or underpaying an employee in Australian conditions due to the high cost of living here. However, in an environment of a lower cost of living, that rate may represent a huge pay raise for that person. It's not so much about trying to achieve the lowest possible hourly rate, it's about boosting efficiency. Some of your best team members in the future may come from these regions.

Locally, you could use a virtual office service to handle incoming mail, or account mail outs if you do that on a monthly basis. You could also have them handle administrative account inquiries perhaps, as opposed to tying up your clinical staff in that process and taking them away from your patients. That could include things like managing accounts payable and receivable, and being able to use modern customer service platforms to be able to achieve this virtually. You can set down your best practices for managing those processes. A self-service portal may assist, as opposed to having to employ someone in accounts to answer phones. I appreciate that this may only be necessary for larger pharmacies and groups of pharmacies, but all of these aspects can be considered for small businesses as well.

Increasing integration

We've spoken a lot about what integrations are possible, and that almost anything is possible. It may be that you've achieved success in some areas of great importance to your business, but now you want to integrate more. To do this, you may need to engage trusted advisers and also your suppliers directly about how they could perhaps have more of their data talk to other partners that you engage with that they don't currently deal with. In terms of monitoring your KPIs and your feedback loop, you may choose to generate reports on a set basis rather than an ad-hoc basis, and you may then choose to find a way of displaying that data as a dashboard, something that you are able to access from a tablet or a smartphone anywhere, anytime, and which can allow you to monitor your business and some of your goals and objectives that you've put in place.

A good technology partner can help you with this, to enable better control and also to minimise the time it takes to find out how you are tracking – you can even get information every hour of every day if you choose. You may be able to set up automated reports with existing technology partners, or there may even be new technology partners who are capable of extracting the data from your current systems and presenting that back to you in a nice report format.

One example of dashboards and reporting might be around increasing generic substitution, which continues to be a top priority in our pharmacies. Our current point-of-sale and dispense systems are simply not capable of giving us live insights into our performance every minute or every hour in our pharmacies. But there are dashboard tools available from some of our product suppliers or generic suppliers that can display this information, and flag each individual patient that may not have been offered generic substitution, or who has turned it down. This enables you to zero in on potential opportunities in your pharmacy and enables your whole team – if you choose – to access this information so they can proactively be working on the goal of increasing generic substitution without you needing to get the data every minute, every hour and feed it back to them. This is incredibly powerful, and can significantly increase your potential for success.

Business intelligence

One way we can continually get better is to know more about our business, and this information all feeds into business intelligence. Introducing more technology into our business has the ability to generate more data and more information, but the information is irrelevant if we cannot piece it together with other data to create knowledge within our business. The step above knowledge is intuition; as business owners we often have a gut feel for something and we run with it, but to have validated knowledge from reliable data is something you may want to consider.

There are business analytical or intelligence tools now available in our industry that allow you to receive custom reports. They match up a bulk set or pool of data from a whole range of systems in your business to give you an impression of how you are tracking based on your goals, expectations and objectives. You may be able to decide on actions based on certain criteria or KPIs hitting certain levels. For example, you could set up a workflow in your business where your systems are aware that the forecast is for a rainy day, and your POS system can adjust your in-store promotional digital displays to show a message around winter and preventing cold and flu. It may even instruct one of your team to place umbrellas at the front of the store.

It's an incredibly powerful thing to consider, but we are only able to achieve this if we open our minds up and decide that in a perfect world we would like all of this to happen. The technology is available, but we need to be thinking big enough to be able to do all of this. It could even be as simple as if the alarm isn't activated and deactivated by a certain time, an automatic text message can be sent to you as the business owner to give your pharmacist in charge a call to see what's going on. Or it may even generate an automatic message to your pharmacist based on who's rostered for that day, as a reminder from you – that you didn't even trigger yourself – to say that you better get a move on and open up the store because you haven't deactivated the alarm yet and it's opening time. Wouldn't that be extremely helpful?

These things can be considered. We are talking about automation and optimisation, and the possibilities are quite endless. Eventually we will have systems thinking like us (for example, artificial intelligence) but I think we are a fair way off from that.

Optimising infrastructure

We need to review our infrastructure. We may want to aim for 100% cloud infrastructure so that we take all of the IT hardware management and maintenance away from our stores. I highly recommend that this be a goal for your business, because the future of our pharmacies is to look after our patients, as pharmacists did in the 1800s. There was no information technology at that point in time, so they never had to employ that. To be able to deliver highly relevant patient care, its important that we remove IT management from the store level completely. But – as we discussed in partnership – you can't achieve everything in 12 months. It's an ongoing journey, but you need to set that goal down so you can reach it.

Optimising communications

Our businesses can change and optimise depending on communications enhancement, so when we start to see the NBN and wireless satellite versions of the NBN, it may give us scope to do so much more. We may already be teleconferencing with our patients using the fastest internet speed available, but once we can get super fast it will allow us to do so much more. We will be able to collaborate in a virtual sense with other healthcare professionals. We could be collaborating with doctors, diabetes educators, podiatrists, eye specialists and renal specialists in the one environment. I have no doubt that with enhanced communications we'll all be able to achieve that.

Optimising feedback

We can optimise our team feedback. We can automate feedback forms to go out to our team to enable them to download their thoughts and best ideas at any time. You can also provide a self-service system, or communicate via an internal communication mechanism to be able to have a constant stream of information that enables better collaboration within your business so that team members will be able to talk to each other. It may not always be one on one – it may be complemented by having a network within your organisation where team members can post thoughts and ideas and be able to collaborate around different projects and patients.

We can also enhance our customer and patient feedback. We can build communities around our pharmacy and our healthcare to enable them to communicate among themselves. There may be people with common disease states. You could create a community where first-time mums can feel confident that they have access to useful resources, not only among themselves but via a trusted pharmacist adviser who is mediating the process. It could be diabetics or arthritic patients – they can receive the best of patient feedback as well as a trusted filtering adviser or a pharmacist being part of that community. You can provide self-service access to a knowledge base, our best knowledge in the hands of our patients and customers anywhere they choose to have it, not just when they come in 9 to 5, Monday to Friday. You can also provide access to feedback from us in a flexible manner that suits them – they may not have time to come into the pharmacy to seek advice. They may want to send us a message, or they may prefer to communicate via a secure communication mechanism, or by phone, or request a call from us at a time of their choosing, not ours. Of course all such systems need to be private and secure.

Augmented reality

We now have the capabilities of technologies such as glass-based interfaces where team members can have "augmented reality" put in front of them. Augmented reality could manifest itself in pharmacy by wearing special glasses that include a little screen in the top corner that enables you to receive information that is relevant for that patient. What a great tool this would be to allow you to conference in other colleagues to be able to help you in some of those patient-facing processes. It might be a particular case or a patient that you want to talk to a more experienced team member about. It might be the patient wants to talk to you as the pharmacy owner and allows you to conference in on the problem that they are experiencing, and you can see the vision that they are seeing. You can be right next to them from anywhere in the world! The outcome for our patients is incredible. You can have your most inexperienced pharmacist talking to one of your most complicated patients, and they can draw on the resources of your most experienced pharmacist to provide expert advice to that patient.

Optimising supplier interactions

Our suppliers can have a deep understanding of our business. Quite often we meet with our suppliers and we will have our data in a report format, and they will have their data in a report format, and that's the only time we are ever collaborating around managing a category or a group of products and getting the best outcome. We now have the capability of exposing that information to our supplier partners on an ongoing basis, so that whenever they choose to have a look and review how our stores are going and provide insights and opportunities for improvement, they can do that at a time of their choosing. It doesn't have to be when we choose to make ourselves available for a collaborative meeting.

This also improves the accuracy of the data that's being transported. A lot of suppliers will take wholesaler withdrawal data and we would use our sales data, and it doesn't always correlate. Wouldn't it be great to be able to make that correlation occur before we have a discussion about our business plan with our suppliers each quarter or every year?

Optimising healthcare provider interactions

The same goes for our interactions with healthcare providers. If we are collaborating with healthcare providers around specific patients, the electronic health record should enable collaboration in digital platform. But at the same time, we can keep our providers connected to our pharmacy via communities (or internal digital platforms; for example, intranet or extranet) where we may be focusing on a particular disease state and putting out the latest information regarding medicines, and providing in-service education to these healthcare partners by doing so. You may visit them monthly or quarterly depending on how important they are to your pharmacy and your patients, and they may come to you and do workshops or presentations with your patients as well, but the ability to continue the conversation beyond this is limited by time and availability. Creating a close-knit community so that we are all working together and getting consistent feedback around what we are doing is incredibly important when we are trying to create the best possible patient experience and put them at the centre, as well as ensuring that our partners are doing the same.

Optimising mobility

Increasing or introducing mobility in your pharmacy can provide great benefits. We all marvel at the Apple experience. When we go to see the Apple consultants, there's never a shortage of them – it's like they were waiting for you to come in and talk to them. They're not pressed for time. They don't need to run around and find things for you. Everything is in the palm of their hand: if you choose to buy a product, they can get it and have it brought to wherever you are standing through their device. They are also able to accept payment and email you a receipt right there and then without moving an inch. This is what we need to consider for our patients as well. What would be the best experience that they can have? Let's build it for them! If computers can be made user accessible and the experience of buying a computer made so easy and seamless, healthcare should be a snap.

📱 *Self-service portals*

There are great benefits in introducing or increasing the use of a self-service portal where patients can interact with our pharmacy and request things and action things without us needing to do it for them. The best example of this is how the travel industry now operates. There are so many things that customers can do as far as booking flights and hotels and so forth, all on their own. Patients would love to do the same thing with their healthcare, but they still need guidance. The difference between travel and health is that if you book the wrong flight, you can always get a refund and book another flight. You take the wrong medication, you may not get the same chance. It's important that we offer self-service to our patients, but self-service with assistance. Whenever we offer anything that's digitally facing our customers, we must approach it in the same manner as if a customer was to come into our pharmacy and self select – would we let that patient walk out the door without ever talking to us? No. If they self select a medicine off the shelf, we will have a pharmacy assistant available to ask and answer questions, and if they are bypassed, we have a pharmacy assistant at the point-of-sale who will ensure that the obligations on us as health care professionals are upheld, and that the patients are asked the right questions to ensure they are not going to run themselves into misadventure with the wrong medicine or advice.

With anything in the digital realm, you need to consider how it would be done in a store environment before you go ahead and let someone design it for you in a digital environment. The digital environment needs to represent your pharmacy in exactly the same way. That's why I refer to this as a hybrid model, in that a patient can feel the same accessing you in a digital sense as in a physical sense. It needs to be congruent.

This applies to customer-facing applications and mobile sites as well. If a patient is on the run, they want the capability of being able to order, pay and collect without being intercepted or held up. But this patient still needs to be given options around accessing advice around those purchases that is flexible to them. That may be retaining information and displaying that back to a patient and asking if anything has changed, and still enabling those chronic or repeat purchases to occur without massive intervention, disruption and frustration from our patient's point of view. We need to work toward achieving professional validation and compliance without being perceived as an annoyance in our patients' lives.

The possibilities are limitless...

Our world has so much possibility right now, and you as a pharmacy owner are in a great position to embrace it and to ensure that your patients receive the best possible advice from you, and not from someone else who may want to enter your space. If the 21st-century pharmacy is where you are and you would love your pharmacy to be in the 22nd century then optimisation is for you. It's exciting to consider that the possibilities for our business, our patients, and ourselves are limitless. New innovation every day has the capability of taking our personal and professional freedom to a whole new level.

Link to download the FREE Transpharmation App for your smartphone/tablet:

To help you anywhere/anytime you have the time to build your 21st-century pharmacy I have developed this helpful app which provides access to your workbook, our community, the latest podcasts and videos, webinars, and much, much more…all available to you on your favourite device.

www.transpharmationbook.com/app

If you don't have a smartphone or tablet device (or would like to begin your journey on your desktop) you can access your online workbook and our member community here:

www.transpharmationbook.com/desktop

Making it happen

Congratulations on completing the four steps of Transpharm-ation! You may wish to revisit the steps again when you are implementing your 12-month roadmap in your pharmacy. Up until now I have given you small anecdotes and key examples of practical ways that the method steps can be applied individually, but to holistically tie the four steps together I have prepared the following resources (only available in your online workbook) for you:

Case studies

These demonstrate the practical application of all four steps, and combined will help you find context and bring it all together.

Community

No doubt you will have encountered a moment in your reading when you wished you could ask me a question regarding the content or your application in your business. I am here for you in our community on a daily basis, where I will answer your questions inside the community so that we can all learn collectively from our insights, questions or problems. Should you wish to maintain privacy over your contribution, this is also available for you.

Support resources

Throughout the online workbook you will have noticed a number of materials, links and support options. These will continue to grow to support you on your journey. Please feel free to comment on these or suggest any new resources that have benefited you in any way.

And much, much more…

Select the **Making it Happen** segment of your workbook to access a special video I have prepared to give you an overview of what to do next and how to complete your roadmap to success.

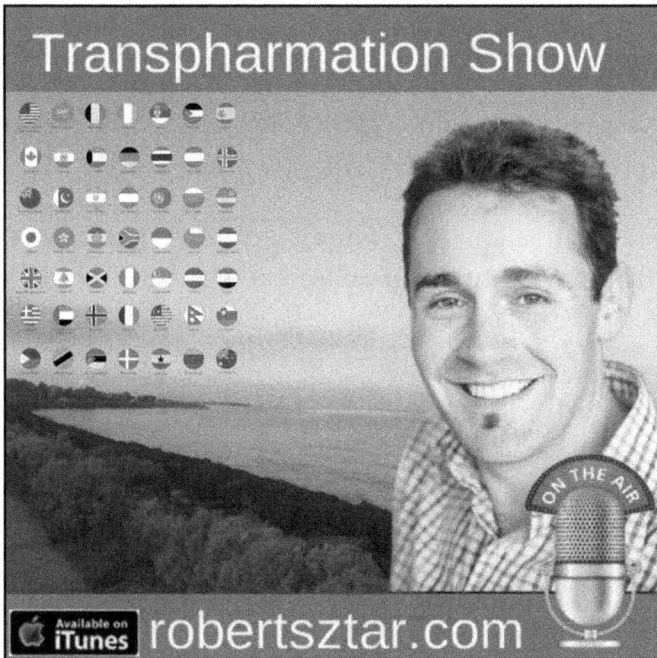
Transpharmation Show
robertsztar.com
Available on iTunes

The **Transpharmation podcast show** was launched in February 2014 to help pharmacy owners build better, smarter, more successful 21st-century businesses by embracing technology. During each episode we cover one or more of the following:

→ listener Q&A (via social media, email, iTunes review, blog comments)

→ current issues and events in technology and pharmacy

→ app of the week

→ a bite-sized DIY transpharmation: this covers a specific technology piece (for example, e-faxing, password management, news feeds, etc.) and how this can benefit the pharmacy owner and their business (time, money and confidence)

→ interviews with successful pharmacy owners who have partnered their business with technology

→ interviews with technology partners who can help pharmacy owners build their 21st-century pharmacy.

The Transpharmation App will alert you to each new episode when it's published, and you can access the entire library from my website robertsztar.com/podcast

Working together

Are you stuck?

My goal as the author of *Transpharmation* is to help guide you on your path to success through the interactive online workbook and provide you with helpful resources and support in the community. However, if following the completion of – or while you are completing – your online workbook you would like more personal, one-on-one assistance that is tailored specifically to you and your business to help you get clarity over any aspect of "Transpharmation", this is available to you:

Click **Working together** inside your online workbook to review the personal support options available.

Business consulting

Are you a company that owns or helps operate multiple pharmacies? I can help you manage multiple businesses using my four-step method, and help you find areas of synergy to boost your group's capabilities in becoming more operationally efficient, implementing a patient-centric business model, and smart use of technology. You can find out more about my consulting services by visiting: www.robertsztar.com/services

Keynote presentations and workshops

If you think other pharmacies or groups of pharmacists would be inspired by the Transpharmation program and its method, please let me know or feel free to pass on my details. I am only too happy to speak or conduct a hands-on workshop at your next conference or group function/meeting on "How pharmacy owners can build smarter, more successful 21st-century businesses before it's too late..."

Please let me know how I can help you.

CONTACT ROBERT

Website: www.robertsztar.com

Email: robert@robertsztar.com

Phone: 1300 798 995 or +61 434 690 579

Twitter: @robertsztar

Join the Transpharmation Community:
www.transpharmation.com.au
and ask Robert a question (he's in there every day!)

About Robert Sztar

For an up-to-the-minute biography and to connect with Robert, please visit his LinkedIn profile:

au.linkedin.com/in/robertsztar/

Robert Sztar is a pharmacist who specialises in introducing pharmacy owners to technology and in partnership help them build their smarter, more successful 21st-century businesses.

He is a second-generation pharmacist, and has been in the industry for 15 years working across community, hospital and international practice settings. For the last seven years, he has had the pleasure of working alongside his wife Amanda (pharmacist) and his father Joseph (pharmacist/pharmacy owner) in Joseph's pharmacies in Victoria, Australia.

The pharmacy industry in Australia has experienced enormous levels of change in the past 10 years. Many of the country's 5500 pharmacies are in a position where the gap between rising operational costs and declining revenue has never been larger, leaving pharmacy owners in survival mode, sacrificing pharmacy staff, barely covering their debts and holding little hope of recovery.

Drawing on his broad experience, Robert discovered that there are three critical elements that when functioning together lead to a genuinely profitable pharmacy:

1. operational efficiency

2. patient-centric business model

3. smart use of technology.

To help pharmacies adopt these elements into their business, Robert and his company Pharmactive have developed a unique four-step method of "Transpharmation", which covers:

1. *Education* – Teaching the purpose of the currently available technologies.

2. *Discovery* – Helping pharmacy owners to discover processes in their business which can be aided or automated by technology.

3. *Partnership* – Introducing pharmacy owners to the right technology partners who will help them to plan, implement, train and maintain their products in their pharmacy.

4. *Optimisation* – Reviewing the solutions implemented to ensure that the original objectives have been delivered upon, and are still relevant in a rapidly changing business landscape.

This is achieved through Robert's book *Transpharmation*, which is positioned as the definitive DIY guide to helping pharmacy owners begin their journey of partnership with technology and start their implementation via an online workbook application (which accompanies the book). This will be supported by a series of mini-transpharmation workshops delivered both in person and online, as well as Robert's flagship 12 to 36 month enablement program of Transpharmation, where he works directly with pharmacy owners to tailor a solution for their individual businesses.

Robert believes that every pharmacy can survive and thrive in the 21st century.

Pharmacy owners working with Robert will be left feeling confident and in control of their 21st-century pharmacy, with more time, more resources, and above all the freedom to choose how they spend their time both in and out of the pharmacy.

"I believe that when you pair a pharmacist's capabilities with technology, it has the power to revolutionise our industry forever."

– Robert Sztar

www.ingramcontent.com/pod-product-compliance
Lightning Source LLC
Chambersburg PA
CBHW070729220326
41598CB00024BA/3360